Springer Proceedings in Business and Economics

Springer Proceedings in Business and Economics brings the most current research presented at conferences and workshops to a global readership. The series features volumes (in electronic and print formats) of selected contributions from conferences in all areas of economics, business, management, and finance. In addition to an overall evaluation by the publisher of the topical interest, scientific quality, and timeliness of each volume, each contribution is refereed to standards comparable to those of leading journals, resulting in authoritative contributions to the respective fields. Springer's production and distribution infrastructure ensures rapid publication and wide circulation of the latest developments in the most compelling and promising areas of research today.

The editorial development of volumes may be managed using Springer's innovative Online Conference Service (OCS), a proven online manuscript management and review system. This system is designed to ensure an efficient timeline for your publication, making Springer Proceedings in Business and Economics the premier series to publish your workshop or conference volume.

More information about this series at http://www.springer.com/series/11960

Panos Pardalos · Ilias Kotsireas ·
Yike Guo · William Knottenbelt
Editors

Mathematical Research for Blockchain Economy

1st International Conference MARBLE 2019, Santorini, Greece

Springer

Editors
Panos Pardalos
Department of Industrial and Systems Engineering
University of Florida
Gainesville, FL, USA

Yike Guo
Data Science Institute
Imperial College London
London, UK

Ilias Kotsireas
Wilfrid Laurier University
Waterloo, ON, Canada

William Knottenbelt
Department of Computing
Imperial College London
London, UK

ISSN 2198-7246 ISSN 2198-7254 (electronic)
Springer Proceedings in Business and Economics
ISBN 978-3-030-37109-8 ISBN 978-3-030-37110-4 (eBook)
https://doi.org/10.1007/978-3-030-37110-4

This Springer imprint is published by the registered company Springer Nature Switzerland AG
The registered company address is: Gewerbestrasse 11, 6330 Cham, Switzerland

MARBLE 2019 Conference Proceedings Volume: Preface

This volume presents the proceedings of the 1st International Conference on Mathematical Research for Blockchain Economy (MARBLE), being held in Santorini, Greece from May 6 to 9, 2019. In contrast to most blockchain conferences and forums which are dedicated to business applications, product development or ICO launches, MARBLE focuses on the mathematics and economics behind blockchain, seeking to bridge the gap between practice and theory. It aims to provide a high-profile, cutting-edge platform for mathematicians, computer scientists and economists to present latest advances and innovations in key theories of blockchain.

The call for papers solicited 24 research papers across a number of themes including incentives, governance, topological analysis, cryptoassets and security. Of the submissions, 15 were accepted for publication and presentation, and of the accepted paper, four of the top-ranked submissions have been chosen for consideration for the Best Paper Award, which will be presented in a special session and voted on by conference attendees. The winner will be announced during the conference banquet, which is to be held in the atmospheric setting of the Volcano Blue restaurant.

The technical programme also features keynotes by the following distinguished speakers: Roman Beck, Jihan Wu, Ambre Soubrian, George Giaglis, Patrick McCorry and Garrick Hileman, and tutorials on the subject of smart contracts (Jerome de Tychey) and zero knowledge proofs (Alexandre Pinto). There is also an industry panel focused on the challenges of blockchain-led transformation of economies, which will be chaired by Naeem Aslam.

We thank all authors who submitted their innovative work to MARBLE 2019 this year. In addition, we thank all members of the Technical Programme Committee and other reviewers, everyone who submitted a paper for consideration, the General Chairs, Prof. Yike Guo and Prof. Panos Pardalos, the Organising Chair Kai Sun, the Local Chair Ilias Kotsireas, the Finance Chair Diana O'Malley, the Publicity Chairs Anna Frankowska and Sam Werner, and members of the Centre for Cryptocurrency Research and Engineering who have contributed in many different ways to the organisation effort, particularly Katerina Koutsouri and

Lewis Gudgeon. Finally, we are grateful to our sponsors, the Brevan Howard Centre for Financial Analysis, Asseth and Aventus.io for their generous support.

The Brevan Howard Centre for Financial Analysis at Imperial College Business School would like to thank all those who were involved and participated in the great success of MARBLE 2019 and is looking forward to supporting and being involved with MARBLE 2020 and future events.

Santorini, Greece
May 2019

Panos Pardalos
Ilias Kotsireas
Yike Guo
William Knottenbelt

Contents

Topological Analysis of Bitcoin's Lightning Network 1
István András Seres, László Gulyás, Dániel A. Nagy and Péter Burcsi

Ping-Pong Governance: Token Locking for Enabling Blockchain Self-governance 13
Paul Merrill, Thomas H. Austin, Justin Rietz and Jon Pearce

Collusion Attack from Hubs in The Blockchain Offline Channel Network 31
Subhasis Thakur and John G. Breslin

Sharing of Encrypted Files in Blockchain Made Simpler 45
S. Sharmila Deva Selvi, Arinjita Paul, Siva Dirisala, Saswata Basu and C. Pandu Rangan

Digital Currencies: A Multivariate GARCH Approach 61
Stamatis Papangelou and Sofia Papadaki

Compact Storage of Superblocks for NIPoPoW Applications 77
Kostis Karantias, Aggelos Kiayias and Dionysis Zindros

On Comparing the Influences of Exogenous Information on Bitcoin Prices and Stock Index Values 93
Luis Montesdeoca and Mahesan Niranjan

Performance of Tip Selection Schemes in DAG Blockchains 101
Richard Gardner, Philipp Reinecke and Katinka Wolter

Committing to Quantum Resistance, Better: A Speed-and-Risk-Configurable Defence for Bitcoin Against a Fast Quantum Computing Attack 117
Dragos I. Ilie, William J. Knottenbelt and Iain D. Stewart

Neural Networks for Cryptocurrency Evaluation and Price Fluctuation Forecasting 133
Emmanouil Christoforou, Ioannis Z. Emiris and Apostolos Florakis

Decentralized Incentive-Compatible and Sybil-Proof Transaction Advertisement 151
Oğuzhan Ersoy, Zekeriya Erkin and Reginald L. Lagendijk

PoolSim: A Discrete-Event Mining Pool Simulation Framework 167
Sam M. Werner and Daniel Perez

Oceanic Games: Centralization Risks and Incentives in Blockchain Mining 183
Nikos Leonardos, Stefanos Leonardos and Georgios Piliouras

Smart Contract-Driven Mechanism Design to Mitigate Information Diffusion in Social Networks 201
Arinjita Paul, Vorapong Suppakitpaisarn and C. Pandu Rangan

Balancing Cryptoassets and Gold: A Weighted-Risk-Contribution Index for the Alternative Asset Space 217
Aikaterini Koutsouri, Francesco Poli, Elise Alfieri, Michael Petch, Walter Distaso and William J. Knottenbelt

An Introduction to the Use of zk-SNARKs in Blockchains 233
Alexandre Miranda Pinto

Topological Analysis of Bitcoin's Lightning Network

István András Seres, László Gulyás, Dániel A. Nagy and Péter Burcsi

Abstract Bitcoin's Lightning Network (LN) is a scalability solution for Bitcoin allowing transactions to be issued with negligible fees and settled instantly at scale. In order to use LN, funds need to be locked in payment channels on the Bitcoin blockchain (Layer-1) for subsequent use in LN (Layer-2). LN is comprised of many payment channels forming a payment channel network. LN's promise is that relatively few payment channels already enable anyone to efficiently, securely and privately route payments across the whole network. In this paper, we quantify the structural properties of LN and argue that LN's current topological properties can be ameliorated in order to improve the security of LN, enabling it to reach its true potential.

Keywords Bitcoin · Lightning network · Network security · Network topology · Payment channel network · Network robustness

Since its launch, Bitcoin [7] gained a huge popularity due to its publicly verifiable, decentralized, permissionless and censorship-resistant nature. This tremendous popularity and increasing interest in Bitcoin pushed its network's throughput to its limits. Without further advancements, the Bitcoin network can only settle 7 transactions per second (tps), while mainstream centralized payment providers such as Visa and Mastercard can process approximately 40,000 tps in peak hours. Moreover one might need to pay large transaction fees on the Bitcoin network, while also need to wait 6 new blocks ($\sim$1 h) to be published in order to be certain enough that the transaction is included in the blockchain.

I. A. Seres (✉) · L. Gulyás · D. A. Nagy · P. Burcsi
Eötvös Loránd University, Budapest 1053, Hungary
e-mail: istvanseres@caesar.elte.hu

L. Gulyás
e-mail: gulya@hps.elte.hu

D. A. Nagy
e-mail: nagy.da@gmail.com

P. Burcsi
e-mail: bupe@inf.elte.hu

P. Pardalos et al. (eds.), *Mathematical Research for Blockchain Economy*,
Springer Proceedings in Business and Economics,
https://doi.org/10.1007/978-3-030-37110-4_1

To alleviate these scalability issues the Lightning Network (LN) was designed in 2016 [8], and launched in 2018, January. The main insight of LN is that transactions can be issued off-blockchain in a trust-minimized manner achieving instant transaction confirmation times with negligible fees, whilst retaining the security of the underlying blockchain. Bidirectional payment channels can be formed on-chain using a construction called Hashed Timelock Contracts (HTLC). Later several payments can take place in a payment channel. The main advantage of payment channels is that one can send and receive thousands of payments with essentially only 2 on-chain transations: the opening and closing channel transactions.

Using these payment channels as building blocks one might establish a payment channel network, where it is not necessary to have direct payment channels between nodes to transact with each other, but they could simply route their payments through other nodes' payment channels. Such a network can be built, because LN achieves payments to be made without any counterparty risk, however efficient and privacy-preserving payment routing remains a challenging algorithmic task [9].

Our contributions. We empirically measure[1] and describe LN's topology and show how robust it is against both random failures and targeted attacks. These findings suggest that LN's topology can be ameliorated in order to achieve its true potential.

1 Background on Lightning Network

In this section we provide a short recap on how LN works. In the following we will use the terms Layer-2 and off-chain interchangeably. LN is a so-called Layer-2 technology, which allows participants issuing transactions without sending a Layer-1 transaction on the Bitcoin parent chain. All parties cooperatively open a channel by locking collateral on the blockchain. The funds can only be released by unanimous agreement or through a pre-defined refund condition [3]. One of the greatest challenge of Layer-2 technologies is to solve how participants can agree on new state updates in a trustless or trust-minimized manner.

Let's take the following toy example: Alice and Bob creates by a single Layer-1 transaction a payment channel with initial balances 10₿ and 0₿ respectively. Straightaway Alice can issue off-chain transactions to Bob up to 10₿. Let's assume Alice issued 3 off-chain transactions to Bob each worth of 1₿. Afterwards Alice's and Bob's balance should be 7₿ and 3₿ and neither Alice, nor Bob should be able to redeem previous balances on the parent chain. LN achieves this by a replace by revocation mechanism, namely both parties collectively authorize a new state before revoking the previous state. Upon dispute, the blockchain provides a time period for parties to prove that the published state is a revoked state [3]. Revoking old channel states is achieved by the exchange of revocation secrets. These secrets, hash preimages, are needed to be retained during the channel's lifetime. A penalty mechanism discourages parties from broadcasting older states. If one party broadcasts a revoked

[1]https://github.com/seresistvanandras/LNTopology.

state, the blockchain accepts within a time-window proofs of maleficence from the other party. A successful dispute allots the winning party *all* coins of the channel. In our example if Alice broadcasts a revoked channel state with balances 8₿ and 2₿, Bob can prove, that Alice maliciously broadcasted a revoked state. The penalty mechanism grants all the 10₿ to Bob.

The great insight of LN, is that if now Alice would like to issue a payment to Cecily, who eventually has already established a payment channel to Bob, then Alice does not need to open a payment channel and create a costly on-chain transaction, rather she can route her payment through her payment channel with Bob to Cecily. However the maximum amount of bitcoins Alice can send to Cecily is the minimum of all the individual balances on the payment route from Alice to Cecily. Hashed time-locked contracts (HTLC) enable routed payments to be atomic. For a technical description of HTLCs and multi-hop payments the astute reader is referred to [8].

In the following we will model LN as an undirected, weighted graph, where nodes are entities who can issue payments using payment channels which are the edges of the LN graph. The weight on the edges, capacities, are the sum of individual balances. Note, that in most cases individual balances are not known to outsiders. Only the capacity of a payment channel is public information, however one can effectively assess individual balances with handy algorithms [4].

2 Lightning Network's Topology

LN can be described as a weighted graph $G = (V, E)$, where V is the set of LN nodes and E is the set of bidirectional payment channels between these nodes. We took a snapshot[2] of LN's topology on the 10th birthday of Bitcoin, 2019 January 3rd. In the following we are going to analyze this dataset. Although the number of nodes and payment channels are permanently changing in LN, we observed through several snapshots that the main topological characteristics (density, average degree, transitivity, nature of degree distribution) remained unchanged. We leave it for future work to analyze the dynamic properties of LN.

LN gradually increased adoption and attraction throughout 2018, which resulted in 3 independent client implementations (c-lightning,[3] eclair[4] and lnd[5]) and 2344 nodes joining LN as of 2019, January 3rd. The density of a graph is defined as $D = \frac{2|E|}{|V||V-1|}$ which is the ratio of present and potential edges. As it is shown in Fig. 1. LN is quite a sparse graph. This is further justified by the fact that LN has 530 bridges, edges which deletion increases the number of connected components. Although LN consists of 2 components, the second component has only 3 nodes. The

[2]https://graph.lndexplorer.com.

[3]https://github.com/ElementsProject/lightning/.

[4]https://github.com/ACINQ/eclair.

[5]https://github.com/lightningnetwork/lnd.

Number of nodes	2344
Number of payment channels	16617
Average degree	7.0891
Connected components	2
Density	0.00605
Total BTC held in LN	543.61855₿
s-metric	0.6878
Maximal independent set	1564
Bridges	530
Diameter	6
Radius	3
Mean shortest path	2.80623
Transitivity	0.1046
Average clustering coefficient	0.304
Degree assortativity	−0.2690

Fig. 1 LN at a glance: basic properties of the LN graph

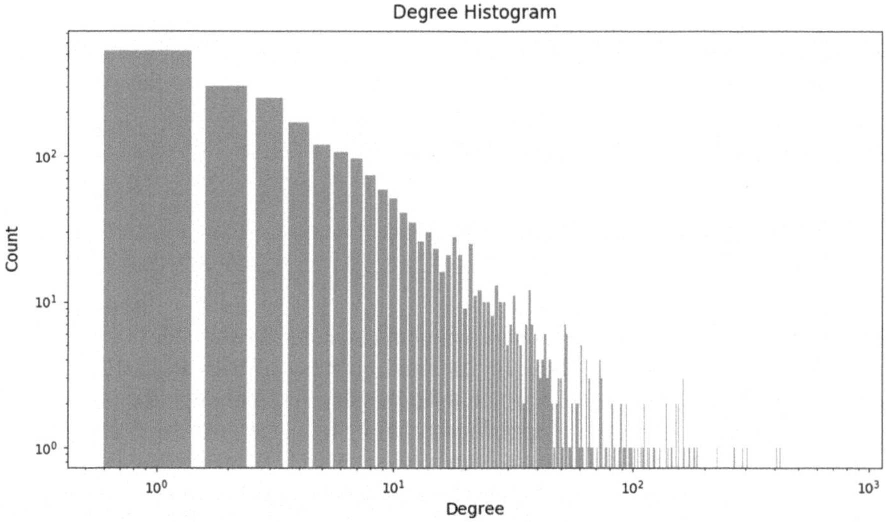

Fig. 2 LN's degree distribution

low transitivity, fraction of present and possible triangles in the graph, highlights the sparseness of LN as well.

Negative degree assortativity of the graph indicates that on average low degree nodes tend to connect to high degree nodes rather than low degree nodes. Such a dissortative property hints a hub and spoke network structure, which is also reflected in the degree distribution, see Fig. 2.

Average shortest path length is 2.80623, without taking into account capacities of edges, which signals that payments can easily be routed with a few hops throughout the whole network. Although this is far from being a straightforward task, since one also needs to take into consideration the capacity of individual payment channels along a candidate path.

When a new node joins LN, it needs to select which other nodes it is trying to connect to. In the *lnd* LN implementation key goals for a node is to optimize its centrality by connecting to central nodes. This phenomena sets up a preferential attachment pattern. Other LN implementations rely on their users to create channels manually, which also most likely leads to users connecting to high-degree nodes. Betweenness centrality of a node v is given by the expression $g(v) = \sum_{s \neq v \neq t} \frac{\sigma_{st}(v)}{\sigma_{st}}$, where σ_{st} is the total number of shortest paths between node s and t, whilst $\sigma_{st}(v)$ is the number of those paths, that pass through v. Closeness centrality of a node v is defined as $CC(u) = \frac{N}{\sum_{u \neq v} d(u,v)}$, where N is the number of nodes in the graph and $d(u, v)$ is the distance between node u and v. Closeness centrality measures how close a node is to all other nodes.

Small-world architectures, like LN, exhibit high clustering with short path lenghts. The appropriate graph theoretic tool to asses clustering is the clustering coefficient [11]. Local clustering coefficient measures how well a node's neighbors are connected to each other, namely how close they are to being a clique. If a node u has $deg(u)$ neighbors, then between these $deg(u)$ neighbors could be at maximum $\frac{1}{2}deg(u)(deg(u) - 1)$ edges. If $N(u)$ denotes the set of u's neighbors, then the local clustering coefficient is defined as $C(u) = \frac{2|(v,w):v,w \in N(u) \wedge (v,w) \in E|}{deg(u)(deg(u)-1)}$.

LN's local clustering coefficient distribution suggestively captures that LN is essentially comprised of a small central clique and a loosely connected periphery.

2.1 Analysis of LN's Degree Distribution

LN might exhibit scale-free properties as the s-metric suggests. S-metric was first introduced by Lun Li et al. in [5] and defined as $s(G) = \sum_{(u,v) \in E} deg(u)deg(v)$. The closer to 1 s-metric of G is, the more scale-free the network. Diameter and radius of LN suggest that LN is a small world. Somewhat scale-freeness is also exhibited in the degree distribution of LN. Majority of nodes have very few payment channels, although there are a few hubs who have significantly more connections as it can be seen in Fig. 2. The scale-freeness of LN is further justified also by applying the method introduced in [2]. The maximum-likelihood fitting asserted that the best fit for the empirical degree distribution is a power law function with exponent $\gamma = -2.1387$. The goodness-of-fit of the scale-free model was ascertained by the Kolmogorov-Smirnov statistic. We found that the p-value of the scale-free hypothesis is $p = 0.8172$, which is accurate within 0.01. Therefore the scale-free hypothesis is plausible for the empirical degree distribution of LN (Fig. 3).

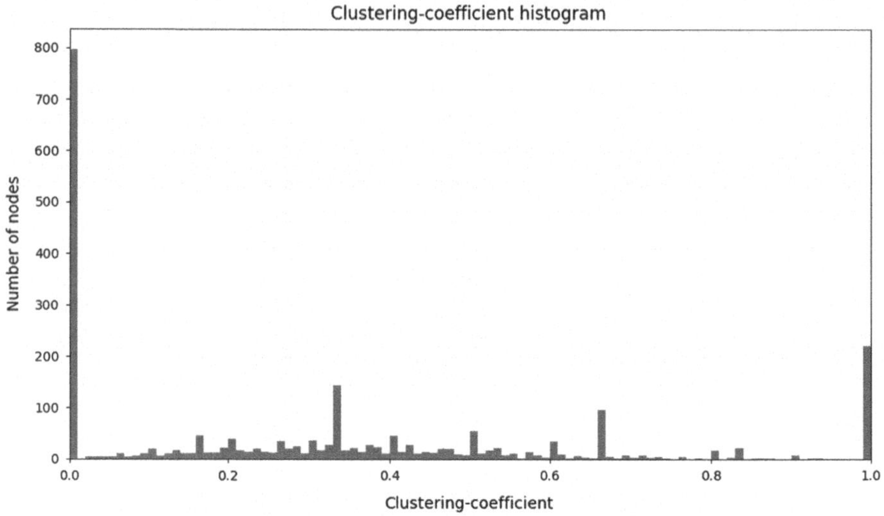

Fig. 3 Local clustering coefficient of LN

3 Robustness of LN

It is a major question in network science how robust a given network is. LN, just like Bitcoin, is a permissionless network, where nodes can join and leave arbitrarily at any point in time. Nodes can also create new payment channels or close them any time. Furthermore as new payments are made, capacities of payment channels are changing steadily. Despite the dynamic nature of LN, its topology's characteristics remain constant after all. In this section we investigate how resilient LN is, whether it can effectively withhold random node failures or deliberate attacks.

Measuring robustness means that one gradually removes nodes and/or edges from the graph, until the giant component is broken into several isolated components. The fraction of nodes that need to be removed from a network to break it into multiple isolated connected components is called the percolation threshold and is denoted as f_c. In real networks percolation threshold can be well estimated by the point where the giant component first drops below 1% of its original size [1].

3.1 Random Failures

Random failures are a realistic attack vector for LN. If nodes happen to be off-line due to bad connections or other reasons, they can not participate in routing payments anymore. Such a failure can be modeled as if a node and its edges are removed from the graph.

Fig. 4 Random failures in networks. Values of critical thresholds for other real networks are taken from [1]

Network	f_c
Internet	0.92
WWW	0.88
US Power Grid	0.61
Mobil Phone Call	0.78
Science collaboration	0.92
E. Coli Metabolism	0.96
Yeast Protein Interactions	0.88
LN	0.96

For scale-free networks with degree distribution $P_k = k^{-\gamma}$, where $2 < \gamma < 3$ the percolation threshold can be calculated by applying the Molloy-Reed criteria, ie. $f_c = 1 - \frac{1}{\frac{\gamma-2}{3-\gamma}k_{min}^{\gamma-2}k_{max}^{3-\gamma}-1}$, where k_{min} and k_{max} denote the lowest and highest degree nodes respectively. This formula yields $f_c = 0.9797$ for LN in case of random failures. This value is indeed close to the percolation threshold measured by network simulation as shown in Fig. 4, that is, LN provides an evidence of topological stability under random failures. In particular this is due to the fact that in LN a randomly selected node is unlikely to affect the network as a whole, since an average node is not significantly contributing to the network's topological connectivity, see also degree distribution at Fig. 2.

3.2 *Targeted Attacks*

Targeted attacks on LN nodes are also a major concern as the short history of LN has already shown it. On 2018 March 21st,[6] 20% of nodes were down due to a Distributed Denial of Service (DDoS) attack against LN nodes. Denial of Service (DoS) attacks are also quite probable by flooding HTLCs. These attack vectors are extremely harmful, especially if they are coordinated well. One might expect that not only state-sponsored attackers will have the resources to attack a small network like LN. In the first attack scenario we removed 30 highest-degree nodes one by one starting with the most well-connected one and gradually withdraw the subsequent high-degree nodes. We recorded the number of connected components. As it is shown in Fig. 5. even just removing the highest-degree node[7] fragments the LN graph into 37 connected components! Altogether the removal of the 30 largest hubs incurs LN to collapse into 424 components, although most of these are isolated vertices. This symptom can be explained by the experienced dissortativity, namely hubs tend to be at the periphery.

We reasserted the targeted attack scenario, but for the second time we only removed one of the 30 largest hubs and recorded the number of connected com-

[6] https://www.trustnodes.com/2018/03/21/lightning-network-ddos-sends-20-nodes.
[7] http://rompert.com/.

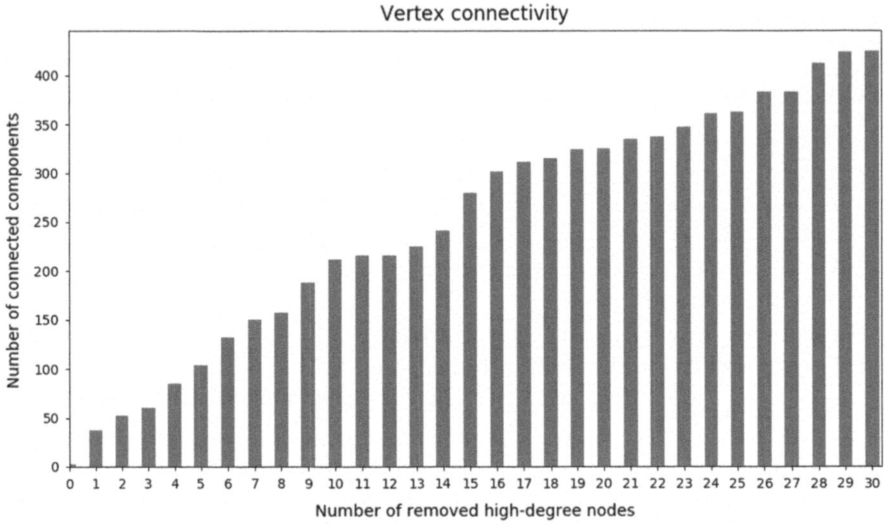

Fig. 5 LN's vertex connectivity, when all the 30 largest hubs are removed one by one

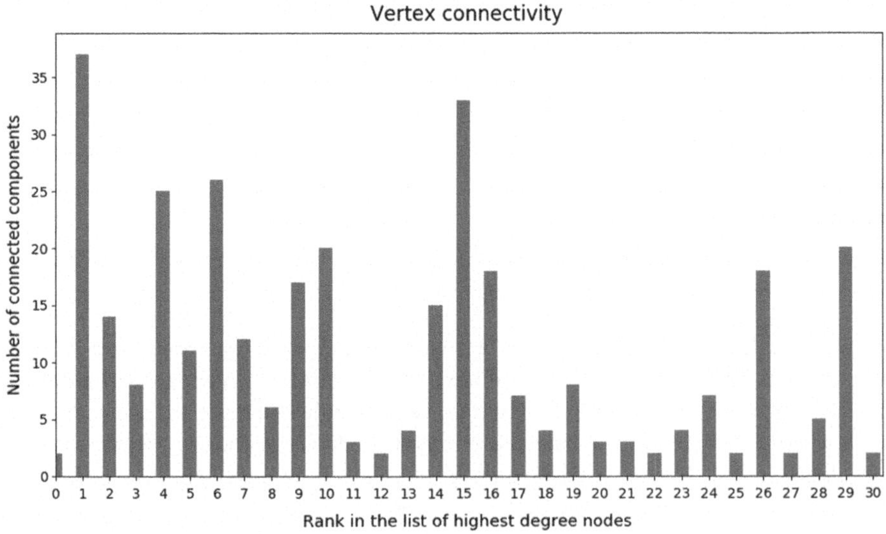

Fig. 6 LN's vertex connectivity if only one high-degree node is removed from the graph

ponents. As it can be seen in Fig. 6 most of the hubs, 25, would leave behind several disconnected components (Fig. 7).

Such network fragmentations are unwanted in case of LN, because they would make payment routing substantially more challenging (one needs to split the payment over several routes) or even impossible (there would be no routes at all).

Fig. 7 Real networks under targeted attacks. Values of critical thresholds for other real networks are taken from [1] to [6]

Network	f_c
Internet	0.16
WWW	0.12
Facebook	0.28
Euroroad	0.59
US Power Grid	0.20
Mobil Phone Call	0.20
Science collaboration	0.27
E. Coli Metabolism	0.49
Yeast Protein Interactions	0.16
LN	0.14

Furthermore we estimated the percolation threshold by simulating two attacking strategies. In the first scenario we removed high degree nodes one by one (high degree removal attack, HDR) and in the second we removed nodes with the highest betweenness centrality (high betweenness removal attack, HBR). Note that in both cases after each node removal we recalculated which node has the highest degree or betweenness centrality in order to have a more powerful attack. We found out that $f_c = 0.1627$ for removing high degree nodes, while for removing high betweenness centrality nodes $f_c = 0.1409$, therefore choosing to remove high betweenness centrality nodes is a better strategy as it can also be seen in Fig. 10.

Node outage not only affects robustness and connectivity properties, but also affects average shortest path lengths and available liquidity. Although the outage of random nodes does not significantly increase the average shortest path lengths in LN, targeted attacks against hubs increase distances between remaining nodes. The spillage of high-degree nodes not only decreases the amount of available liquidity but also rapidly increases the necessary hops along payment routes as Figs. 8 and 9 suggest. This could cause increased ratio of failed payments due to larger payment routes and sparser liquidity. Figure 9 demonstrates how capital allocation is centred upon a few high degree nodes, namely already the removal of as few nodes as 37 decreases the available liquidity by more than 50%. Unfortunately most of these nodes are run by a handful of companies (Fig. 10).

3.3 Improving LN's Resilience Against Random Failures and Attacks

Designing networks which are robust to random failures and targeted attacks appear to be a conflicting desire [1]. For instance a star graph, the simplest hub and spoke network, is resilient to random failures. The removal of any set of spokes does not hurt the connectedness of the main component. However it can not withstand a targeted attack against its central node, since it would leave behind isolated spokes. Furthermore when one attempts increasing robustness of a network, they do desire not to decrease the connectivity of nodes.

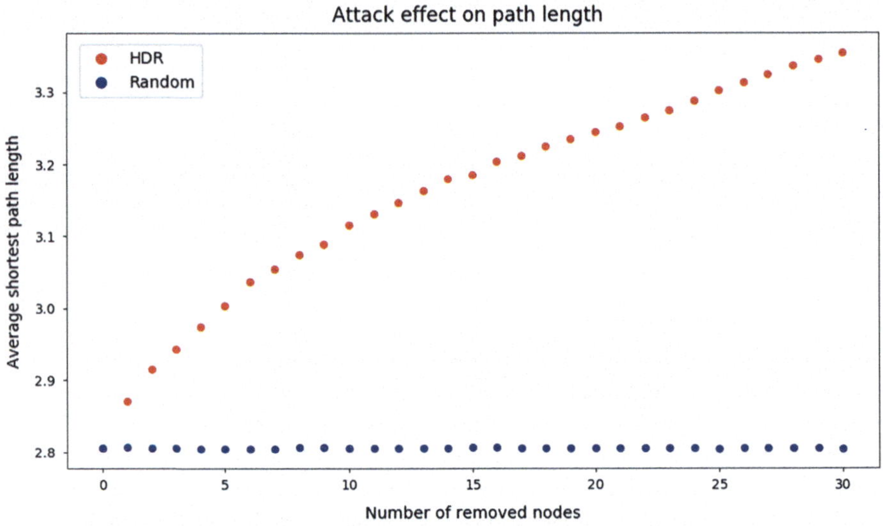

Fig. 8 High degree removal (HDR) attack effects average shortest path lengths

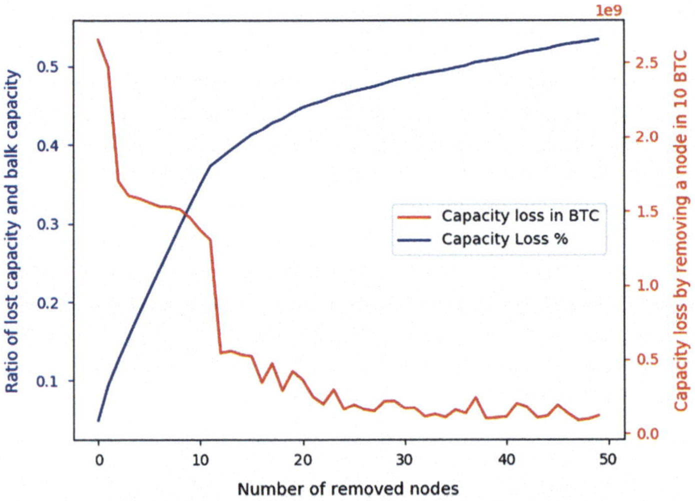

Fig. 9 Lost capacity as removing high-degree nodes

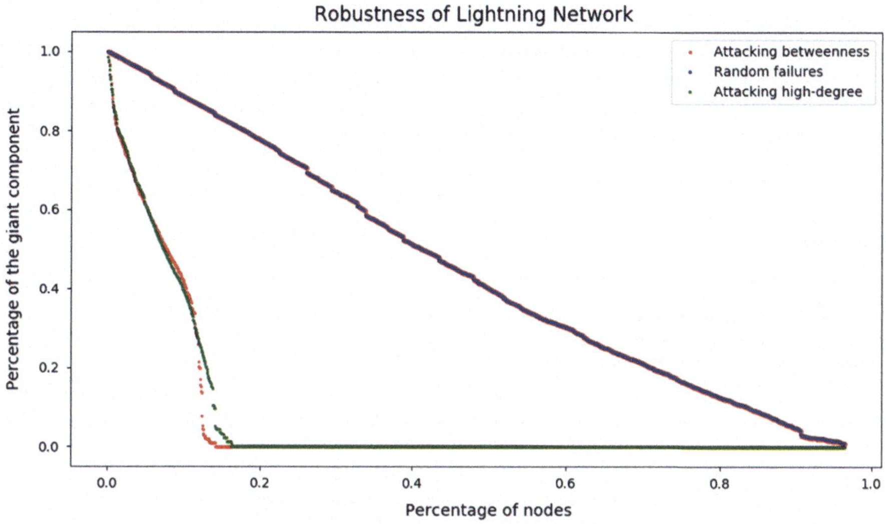

Fig. 10 Percolation thresholds for various attack scenarios: $f_c^{HDR} = 0.1627$, $f_c^{HBR} = 0.1409$, $f_c^{RND} = 0.9645$

A similar optimization strategy of robustness and connectivity to that of [10] could be applied to LN as well. We leave it for future work to empirically assess the robustness and connectivity gains if the strategy of [10] would be implemented in LN client implementations.

Nonetheless, we can still enhance the network's attack tolerance by connecting its peripheral nodes [1]. This could be achieved by LN client implementations by implicitly mandating newcomers to connect to not only hubs, as current implementations do, but also to at least a few random nodes.

4 Conclusion

In summary, a better understanding of the network topology is essential for improving the robustness of complex systems, like LN. Networks' resilience depends on their topology. LN is well approximated by the scale-free model and also its attack tolerance properties are similar to those of scale-free networks; in particular, while LN is robust against random failures, it is quite vulnerable in the face of targeted attacks. High-level depictions of LN's topology convey a false image of security and robustness. As we have demonstrated, LN is structurally weak against rational adversaries. Thus, to provide robust Layer-2 solutions for blockchains, such as LN and Raiden, the community needs to aim at building resilient network topologies.

Acknowledgements We are grateful for the insightful comments and discussions to Chris Buckland, Daniel Goldman, Olaoluwa Osuntokun and Alex Bosworth. Furthermore we are thankful for Altangent Labs (https://www.blocktap.io/) for providing us the invaluable data. The research at Eötvös Loránd University was partially supported by the European Union, co-financed by the European Social Fund (EFOP-3.6.2-16-2017-00012).

References

1. Barabási, A.L., et al.: Network Science. Cambridge University Press (2016)
2. Clauset, A., Shalizi, C.R., Newman, M.E.: Power-law distributions in empirical data. SIAM Rev. **51**(4), 661–703 (2009)
3. Gudgeon, L., Moreno-Sanchez, P., Roos, S., McCorry, P., Gervais, A.: Sok: Off the chain transactions. Cryptology ePrint Archive, Report 2019/360 (2019). https://eprint.iacr.org/2019/360
4. Herrera-Joancomarti, J., Navarro-Arribas, G., Ranchal-Pedrosa, A., Pérez-Sola, C., Garcia-Alfaro, J.: On the difficulty of hiding the balance of lightning network channels. Cryptology ePrint Archive, Report 2019/328 (2019). https://eprint.iacr.org/2019/328
5. Li, L., Alderson, D., Doyle, J.C., Willinger, W.: Towards a theory of scale-free graphs: definition, properties, and implications. Internet Math. **2**(4), 431–523 (2005)
6. Lin, Y., Chen, W., Zhang, Z.: Assessing percolation threshold based on high-order non-backtracking matrices. In: Proceedings of the 26th International Conference on World Wide Web, International World Wide Web Conferences Steering Committee, pp. 223–232 (2017)
7. Nakamoto, S.: Bitcoin: A Peer-to-Peer Electronic Cash System (2008)
8. Poon, J., Dryja, T.: The bitcoin lightning network: scalable off-chain instant payments (2016). https://lightning.network/lightning-network-paper.pdf
9. Roos, S., Moreno-Sanchez, P., Kate, A., Goldberg, I.: Settling payments fast and private: efficient decentralized routing for path-based transactions (2017). arXiv preprint arXiv:1709.05748
10. Shargel, B., Sayama, H., Epstein, I.R., Bar-Yam, Y.: Optimization of robustness and connectivity in complex networks. Phys. Rev. Lett. **90**(6), 068701 (2003)
11. Watts, D.J., Strogatz, S.H.: Collective dynamics of 'small-world' networks. Nature **393**(6684), 440 (1998)

Ping-Pong Governance: Token Locking for Enabling Blockchain Self-governance

Paul Merrill, Thomas H. Austin, Justin Rietz and Jon Pearce

Abstract Updating blockchain-based protocols remains a significant challenge. If the community does not come to an agreement, a *hard-fork* can occur, splitting the blockchain's community. Previous protocols have provided mechanisms to establish community consensus through the protocol itself, but these protocols either facilitate substantial, infrequent updates, or they allow more frequent but only minor changes. This work offers a mechanism that allows clients to vote by locking tokens, making the clients' tokens temporarily unavailable in exchange for their vote. This design introduces an economic cost to voting, allowing us to measure both breadth and depth of support. Since there is an economic cost to voting, we wish to make non-contentious issues cheap to pass, but still allow the community to establish agreement on larger, more disputatious proposals. We achieve this property by a *ping-pong governance model*. An issue is tentatively accepted when it achieves enough votes within a fixed period. The proposal then enters a review period, where the opponents must gather enough votes to veto it. The supporters then have their own opportunity to overrule the veto. This process continues with new voting rounds until one side is unable to exceed the needed threshold, settling the issue. Our simulations show that this model allows the community to come to agreement quickly on popular changes, but still come to resolution when the community is more divided. Finally, we define the ideal properties of a blockchain governance protocol, and evaluate different governance protocols under these criteria.

Keywords Cryptocurrencies · Governance · Token economics

P. Merrill · T. H. Austin
0Chain LLC, Cupertino, USA

P. Merrill · T. H. Austin (✉) · J. Rietz · J. Pearce
San José State University, San José, USA
e-mail: thomas.austin@sjsu.edu

P. Pardalos et al. (eds.), *Mathematical Research for Blockchain Economy*,
Springer Proceedings in Business and Economics,
https://doi.org/10.1007/978-3-030-37110-4_2

1 Introduction

Bitcoin [14] introduced the blockchain and revolutionized distributed applications. A tremendous number of innovations have been built on this concept: Ethereum [21] introduced a Turing-complete programming language on the blockchain; Blockstack [1] showed how it could be used to build a public-key infrastructure; several others [5, 13, 19, 20] leveraged the blockchain to provide distributed storage.

But blockchain-based protocols are difficult to update once launched. Upgrading requires participants to explicitly opt-in to the new protocol by installing new versions of the software. These *hard forks* are undesirable, since they can require significant manual involvement. Furthermore, in cases where the blockchain's community fails to come to an agreement, it can split into two blockchains; the infamous DAO debacle [10] that split Ethereum and Ethereum Classic is perhaps the best known of these.

Governance in the context of blockchain refers to how a blockchain's community can make changes that involve some amount of human interaction [4]. While governance protocols cannot guarantee that the blockchain will not fork over a particularly contentious issue like the DAO debacle, they can provide an authoritative community decision that automates much of the process in order to minimize the work requiring human intervention. We discuss the desired properties of governance protocols in Sect. 2.

Other protocols have attempted to offer a smoother process for the protocol to evolve. Tezos [6] offers one extreme, where an entirely new code base can be proposed. However, due to the potential significance of changes, they can happen only infrequently, and require substantial involvement from the community. Other protocols such as Cypherium [7] offer more limited, but more frequent changes. In the case of Cypherium specifically, changes are limited to minor, 1% configuration adjustments.

In this paper, we seek to provide a voting mechanism that can allow minor and popular changes to happen quickly, but that will also allow more substantial (and potentially controversial) changes.

Our design builds on the token-locking reward model developed by Merrill et al. [12]. In this model, clients who own tokens may temporarily lock them to produce token rewards for service providers. Clients may instead choose to spend their tokens, which produces greater rewards for fewer tokens. However, clients who choose this option lose the ability to re-use their tokens, giving up the "free" aspect of the token-locking reward model.

We leverage this token-locking model to serve as a mechanism for the community to vote on changes. Token rewards are treated as votes; by locking more tokens, clients may dedicate more votes to a proposal that they favor. Similarly, they may allocate token rewards against any proposals that they oppose. A crucial aspect of our design is that our voting mechanism can measure both whether a proposal has broad community support and the degree of support or opposition from different parties for a specific proposal. In extreme cases, smaller stake holders can choose to spend their tokens to increase their voting power. In this manner, a group of clients with a relatively small amount of stake (but with strong support for an initiative) have some hope of defeating opponents with less stake but who are not as strongly motivated.

One concern in voting protocols is that a wealthy supporter of a proposal could sweep in during the last moment of voting to pass a measure before other members of the community could react. This is a particular concern for our protocol, since clients have an economic cost associated with voting for a proposal. Our design addresses this issue by having multiple rounds of voting. If a proposal passes, it is followed by a review period where the community may veto the proposal. If a proposal if vetoed, the community may vote to override the veto. The override itself may be vetoed, which may also be overridden, and so on. To refer to vetoing and overriding generically, we say that a vote is *overturned*. Eventually, one side or the other will exceed a threshold that the other side cannot match and the issue will be settled. To minimize the back and forth votes, the overturning faction must exceed the threshold by a fixed buffer amount.

We refer to our model as the *ping-pong governance model*. Each side must gather enough votes to exceed the threshold to overturn a decision, akin to hitting a ping-pong ball over the net. In ping-pong, the game ends when one side fails to get the ball over the net; in our protocol, the voting ends when one side fails to gather sufficient votes to exceed the overturn threshold.

This paper is organized as follows: Sect. 2 defines the ideal properties of a blockchain governance protocol; Sect. 3 reviews the design of the token-locking reward model, which is the mechanism powering our voting model; Sect. 4 reviews the voting mechanism and the types of changes that we wish to allow; Sect. 5 walks through an example of a proposal going through several rounds of voting; Sect. 6 presents our experimental evidence; Sect. 7 discusses our design; Sect. 8 reviews other blockchain governance protocols; and Sect. 9 concludes.

2 Ideal Properties of a Blockchain Governance Protocol

Different governance protocols for blockchains have different priorities. In this section, we define several properties that would be ideal for a blockchain governance protocol. An ideal governance protocol should offer:

1. *Transparency*. All members of the blockchain community should be able to observe the governance process.
2. *Fairness*. What is "fair" depends on the intentions of the blockchain community. In most, voting power should match the stake of the voter.
3. *Automatic enforceability*. Once a decision has been made, all clients on the blockchain should transition to the new model automatically. In other words, the decision of the governance protocol should be enforced automatically.

4. *Flexibility*. A protocol should be able to handle minor tweaks as well as substantial revisions to the blockchain protocol.
5. *Timeliness*. Especially in the case of urgent issues, the governance protocol should be able to come to consensus relatively quickly.

Our own protocol attempts to be transparent, fair, and flexible. We also seek automatic enforceability and timeliness for more minor changes to the protocol. In terms of fairness, we strive to find a balance between the interests of the major stakeholders and the members of our community at large. We believe that introducing an economic cost to voting is an elegant way of balancing the concerns; major stakeholders are more likely to win on most votes, but are unlikely to override the community if the community at large is more passionate about an issue (and willing to suffer a greater economic cost, relative to their total wealth).

Other protocols have different priorities. Tezos [6] offers all of these properties except for timeliness; changes in their ecosystem are deliberately slow, in order to prevent negative changes from happening too quickly. Cypherium [7] prioritizes timeliness and automatic enforceability, at a cost to fairness (only the validator nodes may vote) and flexibility (only minor configuration changes are permitted).

In Sect. 8, we review several blockchain governance protocols and discuss how they align with these properties.

3 Review of Token Locking Reward Model

We first review the token-locking reward model [12], since that is integral to the design of our voting mechanism. The token-locking reward model was designed to allow clients with tokens to write transactions for "free", in the sense that the clients do not permanently lose their tokens.

For comparison, Bitcoin [14] pays miners with two distinct mechanisms. Coinbase transactions create new tokens to reward miners for creating blocks, regardless of what transactions they produce. The new tokens increase the supply of available bitcoins, meaning that the network is paying miners through inflation. Transaction fees are a second mechanism for rewarding miners in the Bitcoin protocol. Each client who wants their transaction included in a block specifies some amount of bitcoin as a reward to the miner for including the transaction. This mechanism also serves to ease congestion, since clients may specify a larger transaction fee if they are more eager for their transaction to be included.

The token-locking reward model combines these two mechanisms. A client *locks* some tokens for a set period (usually 90 days). In exchange, new tokens are immediately created that can be used to offer a reward to miners for including a transaction in a block.

An advantage of this design is that these token rewards can be used to pay a variety of other service providers. For instance, Merrill et al. [12] allow clients to lock tokens to pay storage providers, referred to as *blobbers*. This design stands in contrast to other protocols such as Tendermint [9], where miners can lock tokens to produce blocks and gain token rewards, but where the mechanism is not designed to pay any other parties.

In this sense, voting can be thought of as an additional service. By locking tokens, clients can generate "rewards" of votes for a proposal. This design introduces an economic cost to voting, which therefore serves as a mechanism for measuring a client's commitment to an issue. In other words, we expect that disinterested clients are likely to lock very few tokens, whereas interested clients will lock more. Section 4 reviews this design in greater detail.

3.1 Transitioning Between Free and Pay Model

In some cases, clients might wish to offer greater rewards than they are able to generate by locking tokens alone. Therefore, the token-locking reward model also allows clients to spend tokens as well.

To facilitate a smooth transition between free and pay models, a client can specify some combination of tokens to be locked and of tokens to be given to the service provider. Since token rewards are paid immediately, the service provider's calculation remains simple. For instance, a miner will select the transactions that offer the highest reward, regardless of whether the rewards were generated by locking tokens or simply given as a reward.

The total reward (R) offered to miners is determined by the interest rate multiplier (M) and the amount of tokens that the client offers to lock (A_{lock}). Additionally, a client may offer to give additional tokens (A_{spend}) to the miner. Giving tokens may be useful when the client wishes to offer a more substantial incentive than they could do by locking tokens alone. The formula for the miner's total reward is given by the formula below:

$$R = (M * A_{lock}) + A_{spend}$$

After the locking period has ended, the client regains the tokens they locked and can use them again, but their spent tokens are gone forever.

3.2 Generating Interest and Staking Tokens

One issue with this reward model is that clients might create "faux services" as a way of generating interest for themselves. To avoid this problem, the reward model legitimizes this behavior—clients may openly lock tokens on the blockchain solely to pay themselves interest.

Service providers may be required to *stake* tokens as well; that is, they might lock tokens which may be sacrificed if they fail to perform their duties. This could make offering a service less attractive, if the service provider was forced to sacrifice the rewards from locking tokens normally. In order to address this point, staking tokens *also* generates rewards. This way, the service provider gains the interest they would have received otherwise, but may earn additional reward for the services that they provide.

Section 4.1 discusses how our governance protocol may be used to burn or seize stake for a under-performing service provider.

4 Voting Mechanism

In this section, we discuss the types of issues that we wish to govern, and we review how the token-locking reward model (Sect. 3) can be used as a tool for voting.

4.1 What We Govern

Our protocol can be used to govern three broad areas: configuration, to adjust the settings of the chain; moderation, to punish bad service providers; and feature proposal, to gauge community interest in deeper revisions to the protocol.

Configuration The community can vote to adjust a number of parameters of the blockchain. Different blockchains will have different parameters to govern; in the case of 0Chain's protocol, these parameters include:

- The interest rate multiplier for generating new tokens.
- The locking period used when generating new tokens.
- Percentages of tokens reserved from the mining fees for third parties. [1]
- The maximum number of miners.
- The voting thresholds.
- The multiplier for "nay" votes, discussed in Sect. 4.4.
- The minimum required tokens to make a proposal.

Moderation Moderation allows the community to police itself. In some proof-of-stake systems, a miner or other entity may stake tokens to ensure good behavior. For instance, miners in Tendermint [9] can lose their stake if they sign multiple blocks at the same height. Our protocol could be used in this manner. For instance, if a service provider is not living up to their terms of service, the community could vote to burn or seize the provider's staked tokens.

A cadre of wealthy stake holders could potentially abuse this mechanism to seize the tokens of less wealthy service providers. To avoid this concern, staked tokens are burned unless they are necessary to make a wounded party whole. For instance, if a client had paid a service provider to store data, those tokens could be used to

[1]Specifically, 0Chain's protocol features *sharders* responsible for the long-term storage of the blockchain, and *validators whose role is to verify that the sharders are storing what they claim.*

pay a different provider to store the same data. An additional concern is that wealthy service providers could attempt to use this mechanism to drive less wealthy providers from the market, though they risk damaging the reputation of the network. Preventing this attack could require some form of attestation, but that is beyond the scope of this work.

We expect that moderation changes are less likely to be controversial and will require a lower threshold level. A party being punished will have a very strong interest in defeating the proposal. However, to defeat the proposal, the service provider would need to dedicate additional coins to defeat the vote, "throwing good money after bad" so to speak, and will therefore only fight it if there is a high chance of overturning the moderation vote. Note that the staked tokens cannot be used for voting, since they are already locked.

Feature Requests Beyond configuration changes or moderation, the community may wish to make more sweeping changes to the blockchain protocol. Following Tezos's approach [6], each proposal contains a hash of the modified code base, eliminating any ambiguity about what changes are proposed. If the proposal is accepted by the community, it will be deployed to testnet to run for 30 days. During that time, clients may vote to veto the change as outlined in Sect. 4.3.

Since these changes are more extensive, we set the threshold for the proposal to a high level, favoring broad community support over flexibility.

Decentralized Application Changes Individual decentralized applications (DApps) will require the same types of changes (configuration, moderation, feature requests) as the blockchain protocol itself. However, since there may be substantially fewer interested parties, each DApp may specify its own threshold settings. The design and challenges are otherwise identical.

4.2 *Proposing Changes*

Now that we have reviewed the types of changes we wish to support, we review the mechanism for proposing these changes. Any client with tokens may make a *proposal* by writing a transaction to the blockchain. We refer to this client as the *champion*.

The process works as follows:

1. A champion writes a transaction to the blockchain, specifying
   - Changes desired.
   - Amount of token rewards to dedicate to this proposal. (To avoid "spam" proposals, a minimum stake can be configured for the blockchain).

This step creates two new *pools*[2]: an *accept pool* that tallies tokens voting in favor, and a *reject pool* that tallies tokens voting to oppose the proposal. We refer to the tokens in the accept pool as *yay votes* and the tokens in the reject pool as *nay votes*.[3]
2. Other clients may add tokens to either pool at any time.
3. Once the yay votes exceed the nay votes by the specified threshold, the proposal tentatively passes. The protocol enters a *review period*, where clients may continue to vote for either side.
4. If the opponents of the *review period* are unable to veto it (Sect. 4.3), then the proposal is accepted and voting is ended. Any configuration changes or moderation actions take effect after 100 blocks. This delay allows the miners to come to consensus on the allowed mining settings, and avoids having to accept multiple valid configurations for the chain. Feature requests are advisory only, and so no action is taken whether they pass or not.

All tokens transferred to the accept and reject pools are unspendable, and can be considered burned. We expect that most clients will vote by locking tokens. However, clients who feel particularly passionate about a change might elect to spend their tokens instead. This design favors the major stake holders (who hold the most tokens), but gives some recourse to the minor stake holders, albeit at a cost of losing future voting power.

4.3 Vetoing Proposals, and Overriding Vetoes

One concern with our design is that a last-minute voter might dump a significant number of votes without giving the opposing side a chance to rally. To avoid this issue, a tentatively accepted proposal enters a review period.

During the review period, opponents of the proposal may add nay votes to the reject pool. If the nay votes exceed the yay votes by 10% before the end of the review period, the proposal is immediately vetoed.

At this point, a new review period begins. The yay votes may override the nay votes. This time, the yay votes must exceed the nay votes by 10% before the end of the review period; if the supporters of the proposal successfully gather enough votes, the veto is overridden, and yet another review period begins immediately.

This process may continue indefinitely. Whenever one side is able to exceed the other by the specified threshold, a new review period begins to *overturn* the previous result; that is, to either veto the proposal or override the veto.

[2]A pool is an account holding locked tokens, where a smart contract dictates the terms for unlocking the tokens. In the case of the accept and reject pools, the tokens are never unlocked.

[3]Token rewards in the reject pool may be weighted differently than tokens in the accept pool. In our discussion of yay and nay votes, the vote tallies are assumed to have been adjusted for their weight already. See Sect. 4.4 for more details.

Once a review period has ended without the result being overturned, the voting is closed; the proposal is either accepted or defeated. Either way, both the proposal pool and the veto pool may be garbage collected after the review period has finished, and any tokens in these pools are effectively burned.

4.4 Weighting of Votes

In our discussion so far, we have treated yay votes and nay votes as having equal weight. However, it may be desirable to weight nay votes more heavily in order to avoid overly controversial changes.

To allow this flexibility, the blockchain has a parameter for a *nay vote multiplier*. For instance, if the nay vote multiplier is set to 4, then while each token added to the accept pool counts as 1 vote, each token added to the reject pool counts as 4 votes.

4.5 Multiple Proposals

It is possible that there could be multiple, simultaneous proposals being voted on at any given time. If proposals conflict on any configuration settings, the most recent proposal accepted overwrites any previous proposals. Should two conflicting proposals be accepted in the same block, the proposal with more votes for acceptance is given priority.

We note that an attacker might propose several proposals in the hope of confusing the opposition about which proposal was the "real" one. However, due to the public nature of the blockchain, the opposition can easily see which proposals have the most votes, and focus their defense on those proposals. Furthermore, every proposal requires a certain minimum amount of stake, depleting the attacker's voting power.

5 Illustrated Example

To illustrate the concepts of our protocol, Fig. 1 shows a proposal in action.

Figure 1a demonstrates the creation of the new proposal. An accept pool and a reject pool are created, with votes allocated to the accept pool by the champion. An "accept threshold" value is set; if this value is reached, the proposal will be tentatively accepted.

In Fig. 1b, enough supporters have committed tokens to the proposal for it to pass, and the changes will be automatically accepted by the network if the proposal is not vetoed within the review period.

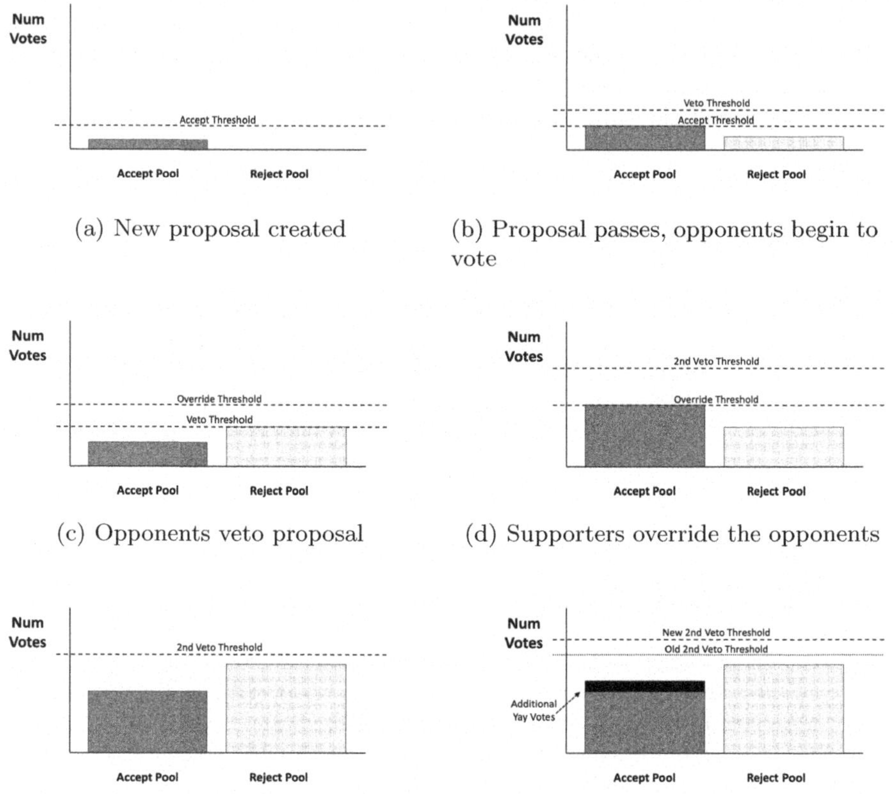

(a) New proposal created

(b) Proposal passes, opponents begin to vote

(c) Opponents veto proposal

(d) Supporters override the opponents

(e) Opponents of the proposal rally

(f) Supporters add votes, veto threshold changes

Fig. 1 Proposal example

In Fig. 1c, enough members find the change to be controversial to collect more 10% more nay votes than yay votes. The proposal is vetoed, and a new review period begins.

Enough members of the community support the proposal to override the nay votes, shown in Fig. 1d. Another review period begins.

The opponents of the proposal rally and commit additional tokens to the reject pool, as shown in Fig. 1e. However, before the opponents of the proposal can gather enough votes, supporters of the proposal allocate additional yay votes to push the veto threshold out of reach of the proposal's opponents, shown in Fig. 1f. Even if the opponents manage to gather enough votes to pass the old second veto threshold, the proposal will be accepted at the end of the review period unless the new second veto threshold is reached.

Once a review period completes without being defeated by the opposition, both the proposal pool and veto pool may be garbage collected. The tokens in these pools are burned.

6 Experimental Results

To validate our approach, we simulated several rounds of voting under different conditions. For these simulations, we used NetLogo v. 6.04.

6.1 Simulating the Speed of Community Agreement

The first question that we are interested in is how quickly the community can come to consensus on an issue depending on how controversial it is.

Figure 2 shows an example of the simulation in action. Clients are represented by either green or red people, depending on whether they support (green) or oppose (red) the proposal, with their shade indicating their degree of support or opposition. The clients relative wealth is determined by their size, ranging from 0 to 1000 tokens.

Blue squares represent "voting booths". Clients wander over the map and enter a voting booth when they land on a blue square. In this manner, voting booths serve to represent the probability of clients voting; more blue squares indicate a higher

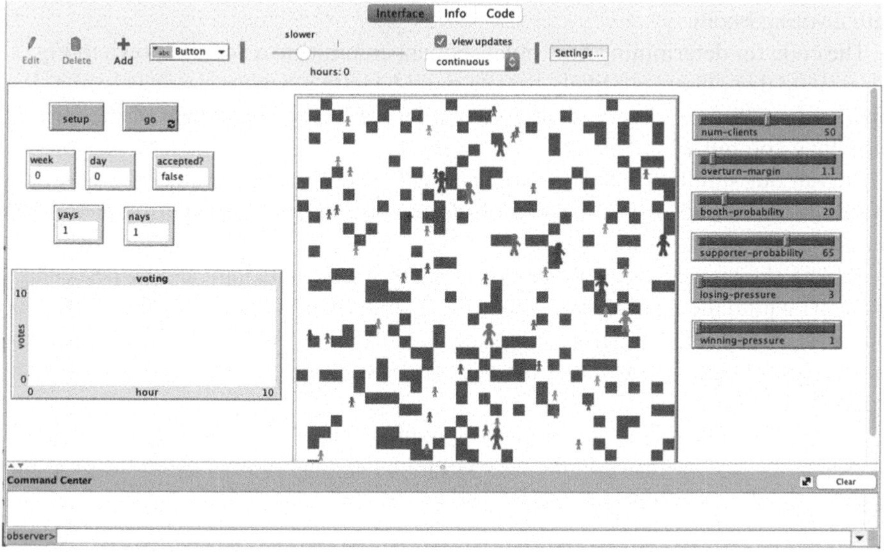

Fig. 2 A sample voting simulation

```
to vote
  if 0 < tokens
  [
    let num-tokens floor(pressure * tokens)
    ; try to create needed tokens by locking
    ; a larger amount for a period of time:
    if num-tokens < (tokens * lock-rate)
    [
      lock-tokens floor(num-tokens / lock-rate)
    ]
    ifelse supporter?
    [
      set yays yays + num-tokens
    ]
    [
      set nays nays + num-tokens
    ]
    set tokens tokens - num-tokens
    ifelse tokens < rich-threshold [set size 1][set size 2]
  ]
end
```

Fig. 3 Voting rules for simulation

probability of voting. The odds of a *particular* client voting are tied to the clients degree of support or opposition. The more interested a client is in the proposal, the greater their degree of mobility, and hence the more likely they are to enter a square with a voting booth.

The code for determining how much a client commits to a vote is shown in Fig. 3. We assume that clients are likely to vote more heavily when their side is losing. We model this with a `pressure` variable that increases the percentage of the client's coins they commit when voting.

We ran our simulation for different percentages of the community supporting a proposal: 95, 80, 65%. For each level of support, we ran the experiment for 100 iterations.

Table 1 shows the results of our experiments. The results indicate the percentage of clients within the community supporting an initiative (Supporters), the number of rounds of voting needed before the issue was resolved (Avg. Number of Rounds), and the percentage of times that the supporters were able to successfully pass the proposal.

The results show that the more the community is divided on an issue, the longer it takes to come to consensus. In the 95% column, the community passed the proposal in 2 rounds (the minimum required) in all cases.

Table 1 Voting results

Supporters (%)	Avg. number of rounds	Yays victorious (%)
95	2.00	100
80	2.04	100
65	2.29	94
50	3.54	40

Table 2 Simultaneous proposal elections

	Number of voting periods
First issue resolved	1.72
Second issue resolved	1.99
Third issue resolved	2.00
Fourth issue resolved	2.03
Fifth issue resolved	2.65

6.2 Simulating Multiple Simultaneous Votes

For our next experiment, we simulated the community voting on several unrelated issues simultaneously. In each run of the experiment, we had 25 clients and the initial threshold was set to 50 votes. The probability of a voting booth was set to 15%.

In contrast to the previous experiments, each client was randomly assigned a level of support between 1 and -1 for each proposal.

Table 2 shows our results for this experiments. Despite the competition of several issues, most proposals are resolved quickly, with the first proposal being finalized in under 2 voting periods on average, and the final issue being resolved in under 3 rounds.

7 Discussion of Rational Behavior of Voters

In this section, we discuss the expected behavior of clients in our protocol, assuming that they are rational actors. Assuming that a client wants a proposal to pass[4] based on pecuniary benefits, the client faces a trade off between (1) the positive return from the proposal passing in the client's favor and (2) the direct loss from both tokens spent on voting and the opportunity cost (if the proposal does pass) of not earning a positive return on tokens spent on voting. Thus, the client faces an expected value function:

$$p(h-v)R+(1-p)(h-v) \tag{1}$$

[4]The same analysis would apply to a client voting against a proposal.

where v is the value (not count) of tokens spent on votes, p is the probability the proposal will pass as a function of v taking other clients' votes as given and assuming $p(0) > 0$ (that is, even if the client doesn't cast votes in favor of the proposal, others will), h is the value of current token holdings, and $R > 1$ is the gross return to the client if the proposal passes including any new tokens the client may acquire because of the proposal. In the case in which the proposal does not pass (with probability $1 - p$), we assume there is no negative impact on the client's total token value beyond the value of tokens spent on votes.

In order to determine the optimal amount the client should spend on votes, we maximize (1) with respect to v:

$$v = \frac{p(R-1)+1}{p'(1-R)} + h \tag{2}$$

where p' indicates the derivative of the function p with respect to v, i.e. the marginal impact of v on the probability of the proposal passing.

Unsurprisingly, as both the gross return and the value of the client's initial token holdings increase, the more the client will want to spend on votes in favor of the proposal. As to the impact of the probability of the proposal passing on vote spending, the results are slightly more nuanced. An increase in the probability of the proposal passing will increase the amount the client wants to spend on votes. This is because if the proposal has a low probability of passing, spending tokens on votes will result in reduced token holdings yet still a relatively small probability of earning return R. However, as the marginal return to a vote, p', increases, the less the client would spend on votes, as a relatively small amount spent on votes would have a relatively larger impact on increasing the probability that the proposal passes, and thus the client can conserve on spending.

Whether the impact of p and p' on the optimal choice of v moves in the same direction is dependent on the functional form of p. If p is concave, an increase in v would would increase p and decrease p' and thus the impact would be clearly positive by Eq. (2). However, if p is convex, an increase in v would increase both p and p', and therefore the net impact would be ambiguous on votes.

8 Related Work

Tezos [6] provides a system for upgrading the blockchain client through community consensus, allowing changes to happen automatically and smoothly. However, only one change may be implemented per voting period, and changes must have substantial buy-in from the community. An additional challenge is that consensus on voting is tied directly to stake. While that approach has advantages, it can result in a minority of powerful stakeholders pushing through proposals that are unpopular with more minor stakeholders.

Tezos amendments take place over 90 days, divided into 4 22.5-day periods. Proposal amendments are submitted by hash value of a tarball of proposed changes (in terms of OCaml code). The amendment that received the most approval in the first quarter is put up for a vote. If quorum is met and the amendment receives 80% yay votes, the amendment is put on Tezos' testnet. Stakeholders then vote to approve/reject test net code for main net.

As discussed in Sect. 2, Tezos offers transparency, flexibility, and automatic-enforceability. It also offers fairness, in that all stakeholders are given a chance to vote, and no party has an influence more powerful than their stake.[5] However, timeliness is not a priority for their design.

EOS [4] provides governance as part of their protocol. Their system features 21 block producers (BPs); among their other responsibilities, BPs may vote on any governance issues. Specifically, they have the power to freeze accounts or update the protocol, as long as 15/21 BPs approve the change. While their design is very powerful, the amount of authority given to the BPs has come under criticism [2, 15]. The influence of the BPs seems to conflict with the goal of fairness, and it is unclear how transparent the decision making process of the BPs is.

EOS's governance protocol does seem flexible, as they are able to implement the same types of changes that we do. Their process is also reasonably timely. The BPs must maintain their 15/21 approval for 30 days. After that, the changes are automatically enforced 7 days later. Furthermore, their design has provisions for *emergency changes*, though the rules for these changes are not clearly specified.

Cypherium [7] allows their protocol to make adjustments to their protocol by committee vote of their validator nodes. Their design is more limited, analogous to the configuration changes in our protocol. In order to provide stability for the economy, the changes are limited to 1% adjustments. This approach does not seem designed to handle larger changes, meaning that it does not offer great flexibility. Also, the fact that the committee of validator nodes votes alone calls into question the fairness of their design. On the other hand, their approach does provide automatic enforcement and timeliness, since changes are enacted immediately. We could not find enough details on their governance protocol to evaluate the transparency of their design.

BOScoin [8] offers voting without tying votes to stake through their "congress network". Using homomorphic encryption, they seek to ensure anonymity but still prevent Sybil attacks. Fairness is an important goal of their protocol; in their case, they seek to make sure that all members of the community have an equal voice, regardless of stake. Unfortunately, we were not able to find enough details on their protocol to evaluate them on the other properties.

Trump et al. [17] discuss the need for "anticipatory governance", where challenges can be identified early on before a hard-fork occurs in a protocol.

[5]The Tezos foundation maintained veto power for the initial operation as they evaluated their governance protocol, breaking the fairness property; they have since relinquished that power [16].

In regards to the economics behind spending resources on voting, the most relevant literature is that on rent seeking or contests. A general overview is provided by Corchón [3].

9 Conclusion and Future Work

In this article, we have explored how the token-locking reward model may be used as the mechanism for a voting protocol. Our approach allows us to measure both the overall level of support among stakeholders, but also to capture when a group of stakeholders has a particularly strong objection to (or support for) a proposal. To our knowledge, no other blockchain governance protocol offers the same ability to measure both breadth and depth of support.

Our analysis shows that our mechanism can be used to arrive at consensus quickly on non-controversial issues, but also allows the community to establish broader support when needed.

The proposed voting mechanism in this paper is similar to a first price, open bid auction. Such auctions can result in both bid shaving (bidders not bidding their maximum willingness to pay in order to avoid the "winners curse") and last minute bidding as a strategy to outbid slower bidders. One mechanism for overcoming these problems are second price, sealed bid auctions as described in Mas-Colell et al. [11]. In this type of auction, also known as a Vickrey auction [18], bidders bid against each other, and the winner pays the second highest bid amount. This prevents bid shaving, as bidders prefer to win a bid versus losing up until the point of their maximum willingness to pay (the bidder's value of the item to be won), and yet don't have to worry about only slightly outbidding their competing bidders. A similar mechanism might be used with the proposed voting mechanism; for example, the winning voters might receive a proportional, partial rebate on the tokens spent on voting. This is an area for future research.

References

1. Ali, M., Nelson, J.C., Shea, R., Freedman, M.J.: Blockstack: a global naming and storage system secured by blockchains. In: USENIX Annual Technical Conference. USENIX Association (2016)
2. Blockgenic: A deep dive into EOS governance (2018). https://medium.com/coinmonks/a-deep-dive-into-eos-governance-49e892eeb4a2
3. Corchón, L.C.: The theory of contests: a survey. Rev. Econ. Des. **11**(2), 69–100 (2007)
4. Eos.io technical white paper v2. Tech. rep., EOS.IO (2018)
5. Filecoin: A decentralized storage network. Tech. rep. Protocol Labs (2017)
6. Goodman, L.: Tezos—a self-amending crypto-ledger. Tech. rep, Tezos Foundation (2014)
7. Guo, S.: Cypherium: a scalable and permissionless smart contract platform. Tech. rep., Cypherium

8. Kim, J., Jun, M., Moon, K., Han, H.: Boscoin white paper 2.0. Tech. rep., BlockchainOS Inc. (2018)
9. Kwon, J.: Tendermint: consensus without mining (2014). https://tendermint.com/static/docs/tendermint.pdf
10. Leising, M.: The ether thief. Bloomberg (2017). https://www.bloomberg.com/features/2017-the-ether-thief/
11. Mas-Colell, A., Whinston, M.D., Green, J.R., et al.: Microeconomic Theory, vol. 1. Oxford University Press, New York (1995)
12. Merrill, P., Austin, T.H., Thakker, J., Park, Y., Rietz, J.: Lock and load: a model for free blockchain transactions through token locking. In: IEEE International Conference on Decentralized Applications and Infrastructures (DAPPCON). IEEE (2019) (forthcoming)
13. Miller, A., Juels, A., Shi, E., Parno, B., Katz, J.: Permacoin: repurposing bitcoin work for data preservation. In: Symposium on Security and Privacy. IEEE Computer Society (2014)
14. Nakamoto, S.: Bitcoin: a peer-to-peer electronic cash system (2009). http://www.bitcoin.org/bitcoin.pdf
15. O'Neal, S.: Corrupt governance? what we know about recent EOS scandal. Cointelegraph (2018). https://cointelegraph.com/news/corrupt-governance-what-we-know-about-recent-eos-scandal
16. Tezos foundation will relinquish first-year veto power. Tezos News (2018). https://tezosnews.us/foundation-relinquish-veto-power/
17. Trump, B., Wells, E., Trump, J., Linkov, I.: Cryptocurrency: governance for what was meant to be ungovernable. Environ. Syst. Decis. **38**, 1–5 (2018). https://doi.org/10.1007/s10669-018-9703-8
18. Vickrey, W.: Counterspeculation, auctions, and competitive sealed tenders. J. Financ. **16**(1), 8–37 (1961)
19. Vorick, D., Champine, L.: Sia: simple Decentralized Storage. Nebulous Inc., Tech. rep. (2014)
20. Wilkinson, S., Boshevski, T., Brandoff, J., Prestwich, J., Hall, G., Gerbes, P., Hutchins, P., Pollard, C.: Storj: a Peer-to-Peer Cloud Storage Network. Storj Labs Inc., Tech. rep. (2016)
21. Wood, G.: Ethereum: a secure decentralised generalised transaction ledger (2014). https://gavwood.com/paper.pdf

Collusion Attack from Hubs in The Blockchain Offline Channel Network

Subhasis Thakur and John G. Breslin

Abstract Offline channels can improve the scalability of blockchains by reducing the number of transactions in the blockchain. Offline channels provide Path-Based fund Transfer (PBT) service which allows a pair of peers without a mutual channel to transfer fund between them using paths in the channel network. In PBTs, peers allow a 3rd party to use their channel for fund transfer in exchange for a transfer fee. There are channels in the Bitcoin Lightning network which are designed to collect such PBT transfer fees. An analysis of Bitcoin's Lightning network revealed the existence of hubs or nodes with very high degree in the channel network. There are only 10 nodes who own more than 50% funds in the Lightning network. These nodes are designed to facilitate PBTs among peers with a low degree (number of channels) in exchange for transfer fees. The emergence of hubs in channel network created the possibility of collusion attack on the channel network where a group of hubs deliberately make few channels non-operational to prevent PBTs involving a selected set of hubs (victims of the collusion attack). In this paper, we model such collusion attack using cooperative game theory and using Banzhaf index we classify the vulnerability of the hubs from the collusion attacks. We propose a design principle of the channel network that can decrease the possibility of collusion attacks.

Keywords Offline channels · Blockchain · Collusion · Banzhaf index

1 Introduction

Scalability is a prominent issue in the blockchain. While Mastercard processes 50,000 transactions per second, Bitcoin processes 7 and Ethereum processes 15 transactions per second. Offline channel [1] is a useful tool to improve the scalability of

S. Thakur · J. G. Breslin (✉)
National University of Ireland, Galway, Ireland
e-mail: john.breslin@nuigalway.ie

S. Thakur
e-mail: subhasis.thakur@nuigalway.ie

P. Pardalos et al. (eds.), *Mathematical Research for Blockchain Economy*,
Springer Proceedings in Business and Economics,
https://doi.org/10.1007/978-3-030-37110-4_3

blockchains. A pair of peers only need to broadcast two transactions to open and close a channel between them. A channel (theoretically) supports an infinite number of transactions between them. Channels offer offline Path-Based fund Transfer (PBT) service [2]. A PBT uses a path in the offline channel network for fund transfer between two parties who do not have a channel. Examples of offline channel networks are Lightning Network for Bitcoin, the Raiden Network [3] for Ethereum and SilentWhispers [2] for credit networks.

Peers allow PBT execution through their channels in exchange for small transfer fee. Hence PBT can be a source of revenue. While ordinary peers with limited funds cannot establish a great number of channels for the purpose of generating revenue, there are financial entities (with access to significant funding) who can establish offline channels for the sole purpose of collecting the PBT transfer fees. In Bitcoin Lightning network we witnessed this phenomenon. There are only 10 nodes with control of more than 50% funds available in the Lightning network. These nodes have a very high degree. We refer to these high degree nodes as the hubs.

Hubs can improve the performance of the offline channels by reducing the PBT completion time and improving the success rate of PBT execution. But it brings a new form of collusion attack on the channel network. In a collusion attack, a collusion (a group of hubs) can make few channels non-operational to prevent PBTs among a set of targeted hubs. The targeted hubs are the victim of the collusion attack. The objective of this paper is to investigate how such a collusion attack can be executed in the channel network and develop a mechanism that can lower the possibility of such anattack. We have the following results in this paper:

1. We present a mathematical model of collusion attack on the channel network. We use cooperative game theory and coalitional power index (Banzhaf index) [4, 5] to model collusion attacks. We model a collusion as a coalition and Banzhaf index gives the estimation on the importance of a hub's participation in a collusion attack.
2. We present a model of the likelihood of collusion attack among the hubs in a channel network using Banzhaf indices.
3. We analyze the possibility collusion attack in the Bitcoin's Lightning network. We found that there are 62 nodes who can execute collusion attacks against 90% of their neighbors in the Lightning network.

The paper is organized as follows: In Sect. 2 we discuss related literature, in Sect. 3 we present the collusion attack problem, in Sect. 4 we present a method to evaluate the possibility of collusion attack in a channel network, in Sect. 5 we present a method to lower the possibility of collusion attack, in Sect. 6 we evaluate Bitcoin's Lightning network to evaluate possibility of collusion attacks and we conclude the paper in Sect. 7.

2 Related Literature

In this paper, we study an attack model in the offline channel network. Offline channels are designed to improve the scalability of blockchains. Examples of such developments are as follows: Bitcoin Lightning network was proposed in [1] which allows peers to create and transfer funds among them without frequently updating the blockchain. Similar networks are proposed for Ethereum [3] and credit networks [2]. A privacy-preserving payment method in the credit network was proposed in [6]. Recent advances on the offline channel network are focused on the development of routing protocols for offline channels. Examples of such routing protocols are as follows: A method for anonymous payment to improve privacy in PBT was developed in [7]. Grunspan and Pérez-Marco [8] proposed a decentralised routing algorithm for the channel network.

Current research in the offline channel network for blockchains is focused on developing better routing protocols for balancing the channels, privacy preserving routing and fast routing protocols. But there is a lack of analysis on the collusion attack that hubs in a channel network can orchestrate. In this paper we analyze collusion attacks among the hubdds in a blockchain peer to peer network. The collusion attack is similar to eclipse attack. Heilman et al. [9] analysed eclipse attack [10, 11] on Bitcoin network and it proposed appropriate countermeasures. Nayak et al. [12] analyzed a combination of selfish mining and eclipse attack on blockchain peer to peer network. In this paper, we analyze collusion attack among hubs in the blockchain network instead of analyzing eclipse attack on the entire peer to peer network. We perform such analysis as we observed centralization of the channel network. Our results can characterize the effect of centralisation in the channel network. We will use the Banzhaf index to characterize collusion attacks. Bachrach and Rosenschein [5] analyzed Banzhaf podddwer indices for network flow games and [4] analyzed Banzhaf power indices for network connectivity games. These research have proved that it is NP-hard to compute the Banzhaf index.

3 Collusion Attacks

First, we present an analysis of the Bitcoin's Lightning network to illustrate the existence of hubs in the channel network. Next, we present the model of collusion attack on the channel network.

3.1 Hubs in Bitcoin Lightning Network

We use the Bitcoin Lightning network data [13] to explain the existence of hubs. The dataset has 2810 nodes and 22,596 edges. The average degree of nodes is 16

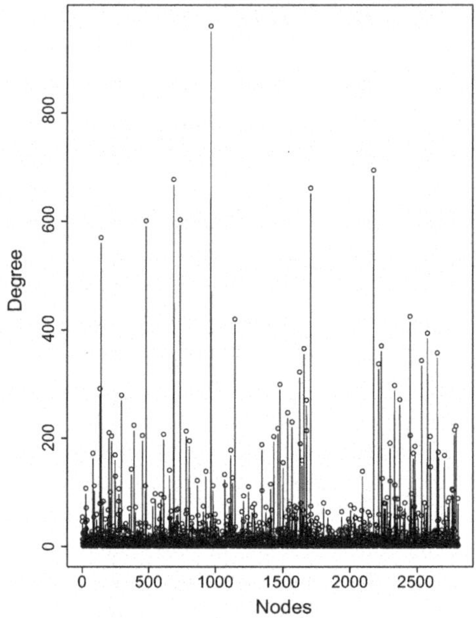

Fig. 1 Degree distribution of bitcoin lightning network data. It shows the existence of very high degree nodes

(shown in Fig. 1). If we consider nodes with a degree more than 50 as hubs then, there are 168 hubs. Collusion is a coalition among the hubs which can prevent PBTs for the remaining hubs. A collusion attack can be executed by creating a cut the channel network.

Collusion is a group of hubs in the channel network who aim to prevent PBTs between a pair of targeted hubs or victims of the collusion attack. We will describe the model of collusion using a neighbourhood of a chosen hub. The neighbourhood will be restricted by the maximum distance from the hub. This will allow us to evaluate the potential of a hub to orchestrate a collusion attack in its neighbourhood. In the next Section we will define such collusion and we will define the potential of a peer to organise collusion as it Banzhaf index. Banzhaf index measures the value of a hub in a coalition (collusion) as it evaluates if the coalition will remain successful (to execute a collusion attack) if this hub leaves the coalition.

3.2 *Models of Collusion Attack*

Let $G = (V, E)$ be a directed graph with n nodes V representing the hubs of the channel network and m edges E representing the channels among the hubs. Let G^i be the subgraph induced by the vertices who are at most k edges apart from $V_i \in V$ (including V_i) on the graph G where k is a positive integer less than diameter of G.

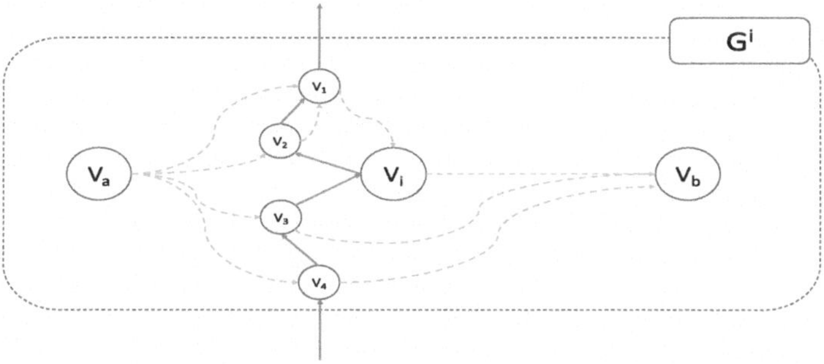

Fig. 2 Example of a collusion centred at V_i that includes V_i, V_1, V_2, V_3 and V_4. The collusion produces a cut between V_a and V_b in G^i

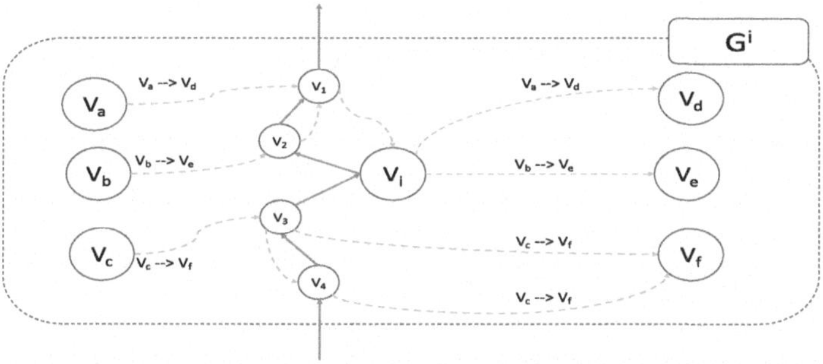

Fig. 3 The collusion is the set of hubs V_1, V_2, V_3, V_4, V_i and the victim of the collusion are the hubs V_a, V_b, V_c, V_d, V_e, V_f. Weight of the collusion is 3 as it disconnects 3 pairs of hubs

V^i will denote the set of nodes at most k edges apart from V_i or the set of vertices of the subgraph G^i.

We will define collusions w.r.t any specific node V_i to use the subgraph G^i. A collusion is a subset of nodes V^i such that:

1. It can produce a cut between a pair of hubs (or more pairs of hubs) in G^i. This pair of hubs is the victim of the collusion attack as PBTs between them will not be executed in G^i.
2. The hubs in the collusion have additional channels to allow the flow of tokens through them.

We formally define collusion (shown in Figs. 2, 3 and 4) as follows:

Definition 1 In a hub network $G = (V, E)$, a collusion C centred at V_i is a subset of V^i such that the following holds:

1. $V_i \in C$.
2. $|C| \leq \delta$ where δ is a positive integer.
3. Let $F \subset E$ be the set of edges originating from any $V_x \in C$.

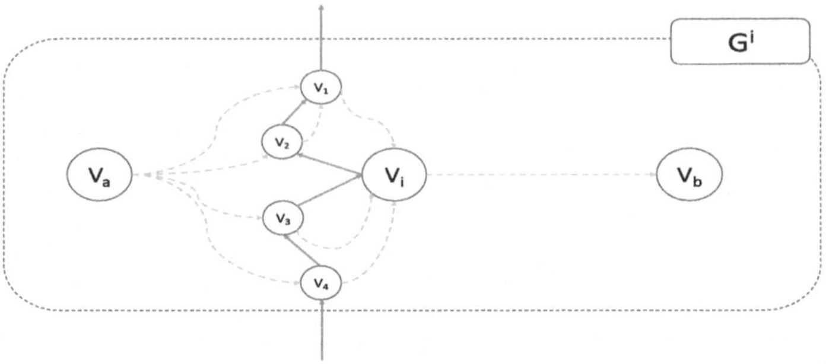

Fig. 4 The collusion is the set of hubs V_1, V_2, V_3, V_4, V_i and the victim of the collusion are the hubs V_a, V_b. V_i is a critical player as the collusion will fail if V_i leaves

4. There exists a pair of nodes $(V_a, V_b) \in V^i - C$ such that there cut $F' \subset F$ where the source is V_a and sink is V_b.
5. There is a path in $F - F'$ that connects every node in C to any node $V_x \in V - V^i$.

The explanation of the above notion of collusion attack is as follows:

1. We define a collusion w.r.t a node V_i. It helps us to define the set of collusions where V_i can have significant contributions.
2. We restrict the size of collusions using the parameter δ.
3. We restrict the size of a subgraph that collusion can control by the parameter k. If collusion can produce a cut between V_a and V_b in the subgraph G^i then it means there is no path in G with distance less than the distance between V_a and V_b in G^i. It means if the collusion blocks the paths between V_a and V_b in G^i then cost of PBT transfer between V_a and V_b is increased by the PBT transfer fee of at least one more channel. Hence a collusion attack can at least increase the cost of PBTs between the victims even if the collusion could not completely prevent any PBTs among its victims.
4. Finally, collusion must have a path to the hubs outside the subgraph G^i despite closing certain channels to execute the collusion attack. It is needed for executing the set of PBTs that the collusion allows.

Definition 2 Weight of a collusion $C \subset V^i$ is the number of pairs of nodes for which the collusion can produce cuts.

Now we define collusion formation game as a cooperative game.

Definition 3 A collusion formation game produces a set of collusions denoted as the set $\{(C, V_i)\}$ where $C \subset V^i$ is the collusion centred at V_i as the result of cooperation among the members of each collusion. The value of a collusion is defined by the function θ as follows:

$$\theta(C) = \begin{cases} 1 & \text{if } C \text{ can produce a cut for a pair of hubs } (V_a, V_b) \in V^i - C \\ 0 & \text{Otherwise} \end{cases} \tag{1}$$

Now we define the importance of a hub in a collusion.

Definition 4 In a collusion (C, V^i), a hub $V_x \in C$ is a critical player if $\theta(C) = 1$ and $\theta(C - V_x) = 0$ indicating that the collusion becomes unsuccessful if V_x leaves the collusion. The number of collusions centred at V_i where V_i is a critical player is denoted by ∇_i.

Now we define the Banzhaf index of a hub for a collusion formation game as follows:

Definition 5 Banzhaf index of the player V_i in the collusion formation game is

$$\beta_i = \frac{\nabla_i}{\sum_{V_x \in V^i} \nabla_x} \tag{2}$$

Note that we restrict the definition of the power of a hub within the subgraph in which it forms collusion. This is because the same subgraph is valid where the hub will be a victim of another collusion attack. Next, we will discuss the algorithm to compute the Banzhaf index.

4 Potential of Collusion Attacks

In this Section, we discuss a method to evaluate the possibility of executing collusion attack in a channel network. First, we will discuss the algorithm to compute Banzhaf index for collusion attack as defined in the previous Section. It should be noted that the computation complexity of computing Banzhaf index is NP-hard [14]. In this paper, we will use Algorithm 1 to estimate Banzhaf indices. The explanation of Algorithm 1 is as follows:

1. In a subgraph G^i centerd at V_i, we compute the number of collusions (subsets of nodes in V_i with maximum cardinality k) where V_i is a critical player.
2. It should be noted that if the number of nodes in G^i is x then the number of collusions where V_i is a member is

$$\frac{x!}{(x-k)!k!} - \frac{(x-1)!}{(x-1-k)!k!} \tag{3}$$

$$\frac{(x-1)!}{(x-1-k)!k!}[\frac{x}{x-k} - 1] \tag{4}$$

The number of possible collusions which includes V_i is very large. It is computationally difficult to test all such collusions to check if V_i is a critical player. Hence instead of checking all collusions we only check the collusions created by a set of random walks from V_i.

Algorithm 1: Computation of Banzhaf index

Data: Hub network as $G = (V, E)$
Result: Banzhaf indices of V as $\{\beta_i\}$
begin
    $Score \leftarrow$ a vector of length n
    for *Each* $V_i \in V$ **do**
        $G^i \leftarrow$ induced subgraph on G by nodes within distance k from V_i
        $d_1 \leftarrow$ degree of V_i in G^i
        $d_2 \leftarrow$ is the diameter of G^i
        $Groups \leftarrow$ a $d_1 \times d_2$
        $N \leftarrow$ neighbours of V_i in G^i
        $j \leftarrow 1$
        for *Each* $V_x \in N$ **do**
            $Groups[j,] \leftarrow$ outcome of a random walk of length k_1
            Remove edges in the path $Groups[j,]$ from G^i
            $j++$
        for *j in [1: size of set Groups]* **do**
            $C \leftarrow j$'th row of the matrix $Groups$
            $H \leftarrow$ created by deleting edges from these vertices with vertices outside C
            $H' \leftarrow$ created by deleting edges from the vertices $C - V_i$ with vertices outside C
            if *Is.connected(H) == FALSE & Is.connected(H') == TRUE* **then**
                $Score[i] \leftarrow Score[i] + 1$
    for *Each* V_i **do**
        $\beta_i = \frac{Score[i]}{\sum_{V_x \in V^i} Score[x]}$

3. For each node V_i we create x random walks where x is the degree of V_i in the graph V^i.
4. For the set of vertices in each such random walk,

   Case 1 We compute the if the deletion of the edges from the set of vertices in each random walk (treated as collusion) to the remaining vertices of V^i (victims of the collusion attack) disconnects the graph G^i.

   Case 2 Next, we compute if such disconnection of the graph is possible without V_i.

5. V_i is a critical player if Case 1 is true and Case 2 is false. Using such information we compute the Banzhaf indices for all hubs.

Now we define the potential of collusion attack in the channel network as follows:

Definition 6 The potential of collusion attack in a channel network can be estimated by the standard deviation of Banzhaf indices of hubs. High standard deviation indicates that there are few hubs who can easily execute collusion attack while the remaining hubs are unlikely to execute collusion attack.

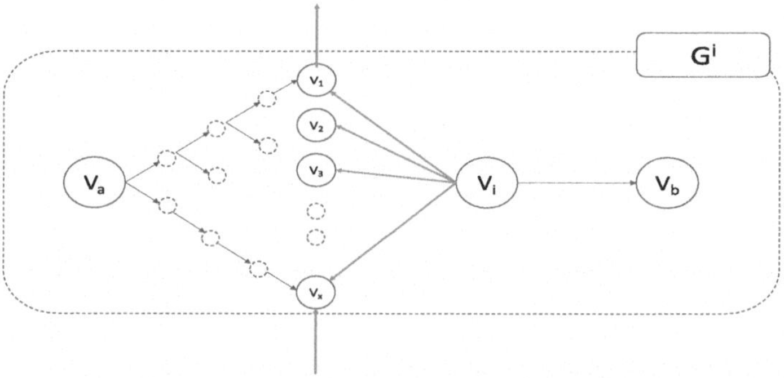

Fig. 5 Relation between degree of an attacker and probability of successful attack: it shows the worst case scenario for V_i as a critical player to execute a collusion attack against V_a and V_b

5 Method to Reduce Possibility of Collusion Attacks

Theorem 1 *The probability that a hub can successfully execute a collusion attack increases as its degree increases.*

Proof Let the collusion C is the set $V_i \cup (V_1, V_2, \ldots, V_x)$ and it wants to execute a collusion attack against the pair of hubs (V_a, V_b) in the subgraph G^i. The attack scenario is illustrated in Fig. 5. Let V_a is at a distance $k - 1$ from V_i and V_b is adjacent to V_i. In this scenario, we will estimate the size of collusion needed to execute an attack against the pair of hubs (V_a, V_b) by V_i. As shown in Fig. 5 the set of collusion is $V_1, \ldots, V_x$. In the worst case, the number of such collusion is the number of leaf nodes of a tree from V_a with depth $k - 2$. Hence the number of nodes is

$$X = 1 + d + d^2 + \cdots + d^{k-2} = \frac{d^{k-2} - 1}{d - 1} \quad (5)$$

where d is the average degree of the hub network. Hence in the worst case V_i needs cooperation from $\frac{d^{k-2}-1}{d-1}$ hubs to attack the pair (V_a, V_b). Note that, V_i can have d_i neighbours (degree of V_i in G^i). The relation between degree of V_i and the probability that V_i can attack on the pair (V_a, V_b) is as follows:

1. V_i may execute the attack if $d_i < \frac{d^{k-2}-1}{d-1}$. It means if V_i has sufficient number of neighbours to form collusion then it can orchestrate such an attack.
2. The probability that V_i has can execute the attack depends on the probability that each node V_1 to V_x has V_i as its neighbour. The probability that the node V_1 is a neighbour of V_i is $d_i \frac{d-1}{d^k-1}$ where $\frac{d^k-1}{d-1}$ is the estimated number of edges in G^i.
3. Hence the probability that V_i can execute the attack is $[d_i \frac{d-1}{d^k-1}]^X$
4. Thus the probability that V_i can successfully attack the pair of hubs (V_a, V_b) increases with the degree of V_i.

Theorem 2 *If hubs of the hub network have a uniform degree then, Banzhaf indices are approximately equal.*

Proof Note that Banzhaf index of a hub depends on the number of collusions where it is a critical player. As proved in the previous theorem, higher the degree higher the probability that a hub can successfully execute a collusion attack. It proves that if the degree of nodes is equal then they will be equally likely to execute successful collusion attacks. Hence their Banzhaf indices will be approximately equal.

We propose that uniform Banzhaf indices may prevent collusion attacks in the hub network. This claim is based on the following observations:

1. In order to detect collusion attack, a hub must observe where its PBT requests are denied in the network. If the network is synchronous then it is a trivial problem. But in an asynchronous network, such detection problem is non-trivial. The collusion detection problem can be formulated as the problem of finding black holes in the network. Block holes are the nodes in a network who destroy mobile agent visiting the node. The collusion detection problem can be formulated as a block hole search problem where mobile agents are network probes. The complexity of this search problem is NP-hard [15, 16]. But several approximation algorithms exist for both synchronous and asynchronous networks [17].
2. If the Banzhaf indices are approximately equal then it means if hub V_a can attack hub V_b then V_b can also execute a collusion attack against V_a.
3. Hence equal Banzhaf indices will bring an 'equilibrium' in the sense that if V_a attacks V_b then V_a can reciprocate such action. Hence it will prevent the hubs from orchestrating collusion attacks.

6 Evaluation with Bitcoin Lightning Network

In this paper, we discussed a model of collusion attack in the offline channels for blockchains. We proved that (a) hubs will have approximately equal Banzhaf indices if their degrees are the same and (b) if hubs have equal Banzhaf indices then they are less likely to initiate a collusion attack. We measure the uniformity of Banzhaf indices as its standard deviation. In this Section, we perform an experimental evaluation of Algorithm 1 and we measure the Banzhaf index of nodes in the Bitcoin Lighting network. We have the following objectives in this experimental evaluation:

1. Prove the correctness of Algorithm 1 which measures the Banzhaf index.
2. Measure the Banzhaf indices of hubs in the Lightning network.
3. Explore the correlation between the uniformity of Banzhaf indices and diameter of a channel network.

We use Bitcoin Lightning network data [13] to analyze collusion attacks. Bitcoin lightning network graph [13] provided an API to access the Lightning network data. The downloaded data is in JSON format and RJSONIO package was used to process the data. The data contains (a) information about each node, i.e., public key and (b) network structure as the edge list. The data was accessed on 1st March 2019. It should be noted that the current size of Lightning network is slightly larger. The data contains the network structure of the Lightning network and it has the following properties:

# Nodes	# Edges	Avg. degree	Min. degree	Max. degree
2810	22,596	16	1	961

In the experimental evaluation, we execute Algorithm 1 using the above data as the input. First, we will evaluate the accuracy of Algorithm 1. We have proved that a peer's Banzhaf index depends on its degree. Greater the degree higher the Banzhaf index. We create a hub network by selecting nodes with a degree in the ranges (30, 100) from the Lightning network data. In this network, we execute Algorithm 1 to estimate the Banzhaf indices of the nodes. The result of such estimation is shown in Fig. 6. It shows that as degree of hubs decreases the Banzhaf index also decreases. Figure 6 provides empirical evidence for the accuracy of Algorithm 1.

Next, we explore the relation between the diameter of a channel network and the uniformity of Banzhaf indices. We generate 17 hub networks from the Lightning network by selecting nodes with minimum degree 30 and maximum degree 50, 55 … 135 respectively. We increase the maximum degree of hub network to increase the diameter of the network (a subgraph of the hub network) as we want

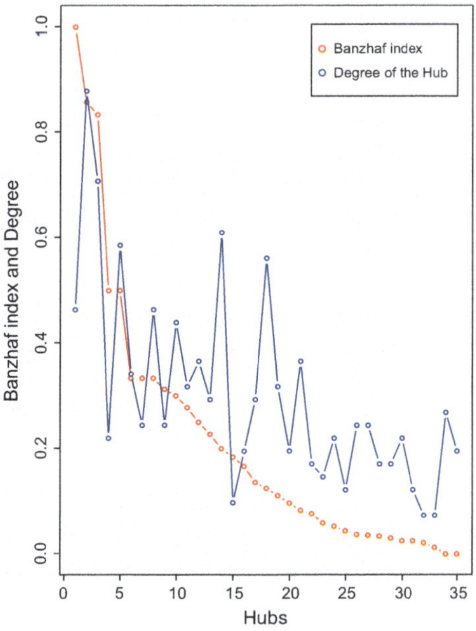

Fig. 6 The Figure shows the relationship between the Banzhaf index and degree of the nodes (normalized to the range [0, 1])

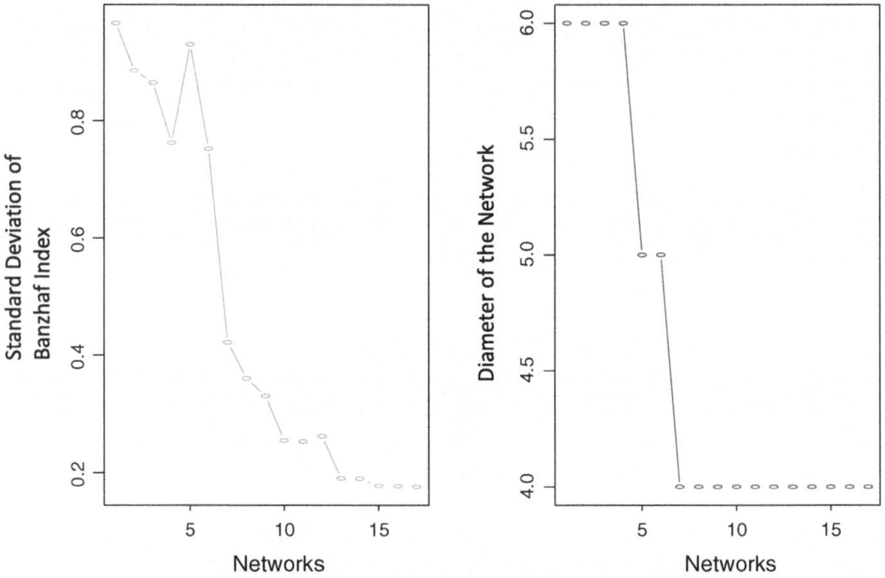

Fig. 7 Relation between the Banzhaf index and diameter of the network. The uniformity of Banzhaf increases as the diameter of the network decreases

to explore the relationship between the Banzhaf index and the network diameter. We use Algorithm 1 to compute Banzhaf indices of all nodes in each such hub network. We observe (shown in Fig. 7) that as we increase the maximum degree of the subgraph generated from the hubs, the diameter of the graph becomes low. As the diameter becomes low it indicates graph converges towards a complete graph. Hence it becomes difficult to generate cuts in such a graph. We keep k (diameter of the collusion graph) as 2 for all datasets. We want to analyze the possibility of collusion within a hub's immediate neighborhood, hence we use diameter 2 because the average diameter of these graphs is 4.5. The computed Banzhaf indices show that power indices become more uniform as the graph evolves towards a complete graph. It means it difficult to execute collusion attacks in channel network if the diameter of the graph becomes small.

Next, we analyze the vulnerability of Bitcoin Lightning network against collusion attack. We use the following metric to measure such vulnerability per node as:

$$\text{vulnerability w.r.t node } v_i = \frac{\text{\# of neighbours with Banzhaf Index less than } v_i}{\text{Size of neighbourhood of } v_i}. \tag{6}$$

We found that there are 62 nodes (shown in Fig. 8) with vulnerability metric at least .9. This means there are 62 nodes who can execute collusion attacks against 90% of their neighbors in the Lightning network.

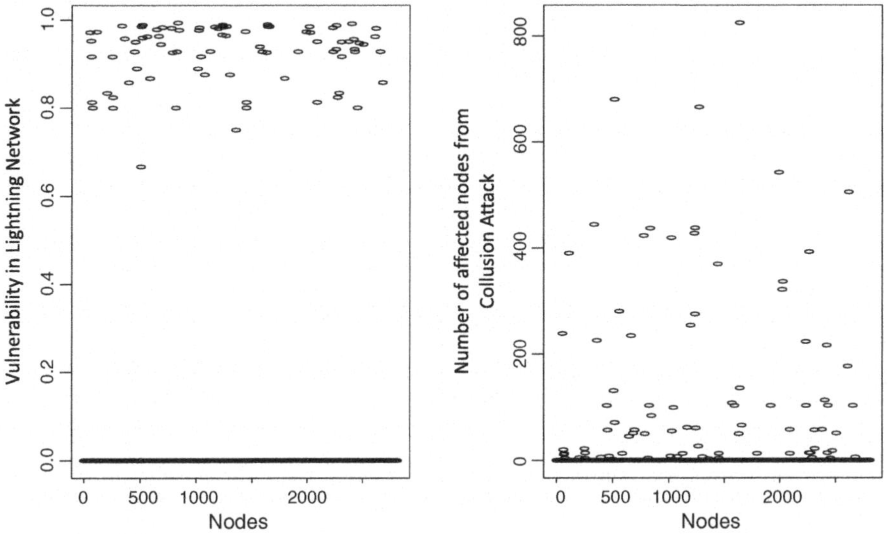

Fig. 8 Vulnerability metric for Lightning network. The left hand figure shows the vulnerability metric for each node in Lightning network and the right hand figure shows the number of neighbours with less Banzhaf index for each node in the Lightning network

7 Conclusion

In this paper, we analyzed collision attacks among the hubs of offline channel networks. We have defined the potential for collusion attacks using Banzhaf indices. We have shown the correlation between uniformity of degree of the hub network and Banzhaf indices. Using experiments on Bitcoin's Lightning network we have shown that as the hub network evolves towards a complete graph it becomes more difficult to create cuts in the graph with a fixed number of edges and hence it increases the uniformity of power indices.

Acknowledgements This publication has emanated from research supported in part by a research grant from Science Foundation Ireland (SFI) and the Department of Agriculture, Food and the Marine on behalf of the Government of Ireland under Grant Number SFI 16/RC/3835 (VistaMilk), co-funded by the European Regional Development Fund.

References

1. Poon, J., Dryja, T.: The bitcoin lightning network: scalable off-chain instant payments. https://lightning.network/lightning-network-paper.pdf
2. Malavolta, G., Moreno-Sanchez, P., Kate, A., Maffei, M.: Silentwhispers: enforcing security and privacy in decentralized credit networks. IACR Cryptol. ePrint Arch. **2016**, 1054 (2016)
3. Raiden Network. http://raiden.network/. Accessed 1 Mar 2019

4. Aziz, H., Lachish, O., Paterson, M., Savani, R.: Power indices in spanning connectivity games. In: Goldberg, A.V., Zhou, Y. (eds.) Algorithmic Aspects in Information and Management, pp. 55–67. Springer, Berlin Heidelberg (2009)
5. Bachrach, Y., Rosenschein, J.S.: Computing the Banzhaf power index in network flow games. In: AAMAS (2007)
6. Moreno-Sanchez, P., Kate, A., Maffei, M., Pecina, K.: Privacy preserving payments in credit networks: enabling trust with privacy in online marketplaces. In: NDSS (2015)
7. Green, M., Miers, I.: Bolt: anonymous payment channels for decentralized currencies. In: Proceedings of the 2017 ACM SIGSAC Conference on Computer and Communications Security, CCS '17, pp. 473–489. ACM, New York, NY, USA (2017). https://doi.org/10.1145/3133956.3134093
8. Grunspan, C., Pérez-Marco, R.: Ant routing algorithm for the lightning network. CoRR (2018). ArXiv:1807.00151
9. Heilman, E., Kendler, A., Zohar, A., Goldberg, S.: Eclipse attacks on bitcoin's peer-to-peer network. In: Proceedings of the 24th USENIX Conference on Security Symposium, SEC'15, USENIX Association, pp. 129–144. Berkeley, CA, USA (2015). http://dl.acm.org/citation.cfm?id=2831143.2831152
10. Castro, M., Druschel, P., Ganesh, A., Rowstron, A., Wallach, D.S.: Secure routing for structured peer-to-peer overlay networks. SIGOPS Oper. Syst. Rev. **36**(SI), 299–314 (2002). https://doi.org/10.1145/844128.844156
11. Singh, A., Ngan, T.W., Druschel, P., Wallach, D.S.: Eclipse attacks on overlay networks: threats and defenses. In: 25th IEEE International Conference on Computer Communications on Proceedings IEEE INFOCOM 2006. pp. 1–12 (2006)
12. Nayak, K., Kumar, S., Miller, A., Shi, E.: Stubborn mining: Generalizing selfish mining and combining with an eclipse attack. In: 2016 IEEE European Symposium on Security and Privacy (EuroS P), pp. 305–320 (2016). https://doi.org/10.1109/EuroSP.2016.32
13. Bitcoin Lightning Network Graph. https://graph.lndexplorer.com/api/graph. Accessed 1 Mar 2019
14. Werman, S., Zohar, A.: Avoiding deadlocks in payment channel networks. In: Garcia-Alfaro, J., Herrera-Joancomartí, J., Livraga, G., Rios, R. (eds.) Data Privacy Management, Cryptocurrencies and Blockchain Technology, pp. 175–187. Springer International Publishing, Cham (2018)
15. Czyzowicz, J., Kowalski, D., Markou, E., Pelc, A.: Searching for a black hole in tree networks. In: Proceedings of the 8th International Conference on Principles of Distributed Systems, pp. 67–80. OPODIS'04. Springer, Berlin, Heidelberg (2005)
16. Czyzowicz, J., Kowalski, D.R., Markou, E., Pelc, A.: Complexity of searching for a black hole. Fundam. Inform. **71**(2–3), 229–242 (2006). http://content.iospress.com/articles/fundamenta-informaticae/fi71-2-3-05
17. Dobrev, S., Flocchini, P., Prencipe, G., Santoro, N.: Mobile search for a black hole in an anonymous ring. Algorithmica **48**(1), 67–90 (2007)

Sharing of Encrypted Files in Blockchain Made Simpler

S. Sharmila Deva Selvi, Arinjita Paul, Siva Dirisala, Saswata Basu and C. Pandu Rangan

Abstract Recently, blockchain technology has attracted much attention of the research community in several domains requiring transparency of data accountability, due to the removal of intermediate trust assumptions from third parties. One such application is enabling file sharing in blockchain enabled distributed cloud storage. Proxy re-encryption is a cryptographic primitive that allows such file sharing by re-encrypting ciphertexts towards legitimate users via semi-trusted proxies, without them learning any information about the underlying message. To facilitate secure data sharing in the distributed cloud, it is essential to construct efficient proxy re-encryption protocols. In this paper, we introduce the notion of proxy self re-encryption (SE-PRE) that is highly efficient, as compared to the existing PRE schemes in the literature. We show that our self encryption scheme is provably CCA secure based on the DLP assumption and our proxy re-encryption scheme with self encryption is CCA secure under the hardness of the Computational Diffie Hellman (CDH) and Discrete Logarithm (DLP) assumption. Our novel encryption scheme, called self encryption, has no exponentiation or costly pairing operation. Even the re-encryption in SE-PRE does not have such operations and this facilitates the service provider with efficiency gain.

S. S. D. Selvi · A. Paul · C. P. Rangan (✉)
Department of Computer Science and Engineering, IIT, Madras, India
e-mail: prangan@cse.iitm.ac.in

S. S. D. Selvi
e-mail: sharmioshin@gmail.com

A. Paul
e-mail: arinjita@cse.iitm.ac.in

S. Dirisala · S. Basu
0chain LLC, San Jose, USA
e-mail: siva@0chain.net

S. Basu
e-mail: saswata@0chain.net

P. Pardalos et al. (eds.), *Mathematical Research for Blockchain Economy*,
Springer Proceedings in Business and Economics,
https://doi.org/10.1007/978-3-030-37110-4_4

1 Introduction

The recent explosion of data volumes and demand for computing resources have prompted individuals and organisations to outsource their storage and computation needs to online data centers, such as cloud storage. While data security is enforced by standard public-key encryption mechanisms in the cloud, secure data sharing is enabled by efficient cryptographic primitives such as proxy re-encryption (PRE). PRE enables re-encryption of ciphertexts from one public key into another via a semi-trusted third party termed *proxy*, who does not learn any information about the underlying plaintext. A user can delegate access to his files by constructing a special key, termed as re-encryption key, using which the proxy performs the ciphertext transformation towards a legitimate delegatee. PRE systems can be classified into unidirectional and bidirectional schemes based on the direction of delegation. They can also be classified into single-hop and multi-hop schemes based on the number of re-encryptions permitted. In this work, we focus on unidirectional and single-hop PRE schemes.

The current model of cloud storage is operated through centralised authorities, which makes such a system susceptible to single point failures and permanent loss of data. Recently, blockchain technology, initially designed as a financial ledger, has attracted the attention of researchers in a wide range of applications requiring accountable computing and auditability. Blockchain enabled distributed peer-to-peer cloud storage solutions are steadily replacing its centralised counterpart. A blockchain provides multiple parties to agree upon transactions and contracts in an immutable and auditable way. Decentralised applications such as dApp providers make use of this capability to provide services that are transacted in a publicly verifiable manner. When the service provided by the dApp is not directly from the dApp owner itself but from other third parties, it brings up additional challenges. How would the end user using the dApp trust that the unknown third party service provides used by the dApp are trust worthy? This issue is specifically addressed, for example, by the dApp called **0box** [1] provided by **0Chain** [2]. Such a storage dApp allows any user to upload and share their files to their friends and families similar to many other popular storage services. However, most existing services trust the service provider and upload the content without any encryption. But 0box strives to provide zero-knowledge storage such that the third party storage providers will not know the uploaded content. This is achieved using an efficient CCA-secure proxy re-encryption scheme, outlined in this paper. When a user shares the encrypted content with a trusted party, he provides the re-encryption keys using the public key of the trusted party so that only that party is able to decrypt the content. By facilitating the associated transactions on the blockchain, this scheme provides end-to-end transparency and security for end users to procure storage services at highly competitive prices without worrying about the reputation of the storage providers. We first propose a novel self-encryption (SE) scheme, which is much more efficient than the standard CPA secure El-Gamal encryption scheme. This work is further extended to design a CCA-secure proxy re-encryption scheme (SE-PRE) that adds re-encryption functionality to self-encryption. Prior to our work, the most efficient PRE construction was reported in [3] by Selvi et al. We show that our PRE design is much more efficient than the scheme in [3].

Proxy Re-encryption (PRE): Proxy re-encryption is a term coined by Blaze, Bleumer, and Strauss [4] and formalized by Ateniese, Fu, Green, and Hohenberger [5, 6]. PRE has been studied extensively for almost two decades [3–7]. A good survey of the PRE schemes and security models of PRE can be found in [8, 9].

2 Preliminaries

In this section we give the definitions of various assumptions adopted for proving the security of the proposed schemes, the general and security model of **SE** and **SE-PRE** schemes.

2.1 Definition

Definition 1 *Discrete Logarithm Problem (DLP)*: The discrete logarithm problem in a cyclic group $\mathbb{G}$ of order q is, given (q, P, Y) such that q is a large prime, P, $Y \in \mathbb{G}$, find $a \in \mathbb{Z}_q^*$ such that $Y = aP$.

Definition 2 *Computation Diffie Hellman Problem (CDH)*: The Computation Diffie Hellman Problem in a cyclic group $\mathbb{G}$ of order q is, given (q, P, aP, bP) such that q is a large prime, P, aP, $bP \in \mathbb{G}$, find Q such that $Q = abP$, where $a \in \mathbb{Z}_q^*$.

2.2 Generic Model of Self-encryption (SE)

The self encryption (**SE**) is a novel primitive that allows an user to store their files securely with minimal computation overhead. This primitive is different from the traditional public key encryption approach as encryption can be done only by the owner of the file who possess the private key related to the public key which is used for encrypting the file. It has the following algorithms:

1. ***Setup*** (κ): *This algorithm is run by the trusted entity. On input of a security parameter* κ, *the* ***Setup*** *algorithm will output the system parameters Params.*
2. ***KeyGen*** $(U_i, Params)$: *This algorithm is run by the user* U_i. *This is used to generate a public and private key pair* (PK_i, SK_i) *for the user* U_i.
3. ***Self-encrypt*** $(m, t_w, SK_i, PK_i, Params)$: *The encryption algorithm is run only by the user* U_i. *This algorithm requires the knowledge of the private key* SK_i *corresponding to the public key* PK_i *of user* U_i. *This takes as input the message* m, *the tag* t_w, *the private key* SK_i *and public key* PK_i *of user* U_i. *It will output a ciphertext* C *which is the encryption of message* m *under the public key* PK_i *and*

tag t_w. This approach differs from the traditional public key encryption where the encrypt algorithm can be run by any user.

4. ***Self-decrypt*** $(C, SK_i, PK_i, Params)$: *The decryption algorithm is run by the user U_i. On input of the ciphertext C, the private key SK_i and the public key PK_i of user U_i, this will output the message m if C is a valid self encryption of m under PK_i, SK_i and t_w. Otherwise, it returns $\perp$.*

2.3 Generic Model of Proxy Re-encryption with Self-encryption (SE-PRE)

The SE-PRE is a proxy re-encryption primitive that uses a self encryption scheme as the base algorithm and provides a mechanism to delegate the self-encrypted ciphertext. The SE-PRE scheme consists of the following algorithms:

1. ***Setup*** (κ): *The setup algorithm takes as input a security parameter κ. This will output the system parameters Params. This algorithm is run by a trusted party.*
2. ***KeyGen*** $(U_i, Params)$: *The key generation algorithm generates a public and private key pair (PK_i, SK_i) of user U_i. This algorithm is run by a user U_i.*
3. ***ReKeyGen*** $(SK_i, PK_i, PK_j, c_w, Params)$: *The re-encryption key generation algorithm takes as input a private key SK_i of delegator U_i, public key PK_i of delegator U_i, public key PK_j of delegatee U_j and condition c_w under which proxy can re-encrypt. It outputs a re-encryption key $RK_{i \to j}$. This is executed by the user U_i.*
4. ***Self-encrypt*** $(m, t_w, SK_i, PK_i, Params)$: *The self encryption algorithm takes as input the message m, the tag t_w, the private key SK_i of user U_i and public key PK_i of the user U_i. It outputs a ciphertext C which is the encryption of message m under the public key PK_i, private key SK_i and tag t_w. This algorithm is executed by the user U_i.*
5. ***Re-encrypt*** $(C, PK_i, PK_j, c_w, RK_{i \to j}, Params)$: *The re-encryption algorithm takes as input a self-encrypted ciphertext C, the delegator's public key PK_i, the delegatee's public key PK_j, the condition c_w and a re-encryption key $RK_{i \to j}$ corresponding to c_w. It outputs a ciphertext D which is the encryption of same m under public key PK_j of user U_j. This is run by a proxy who is provided with the re-encryption key $RK_{i \to j}$.*
6. ***Self-decrypt*** $(C, SK_i, PK_i, Params)$: *The self decryption algorithm is run by the user U_i. This will take as input the ciphertext C, the private key SK_i of user U_i and public key PK_i of user U_i. It will output the message m if C is a valid encryption of m under PK_i and SK_i of user U_i and tag t_w. If C is not valid, this algorithm returns $\perp$.*
7. ***Re-decrypt*** $(D, SK_j, Params)$: *The re-decryption algorithm takes as input a re-encrypted ciphertext D and a private key SK_j of user U_j. It outputs a message $m \in \mathcal{M}$, if D is a valid re-encrypted ciphertext of message m or the error symbol $\perp$ if D is invalid. This algorithm is run by the user U_j.*

2.4 Security Model

In this section we present the security model for the self-encryption scheme and the proxy re-encryption scheme. The security model gives details about the restrictions and oracle accesses given to the adversary. It is modelled as a game between a challenger $\mathcal{C}$ and an adversary $\mathcal{A}$.

2.5 Security Model for Self-encryption

The security of Self-encryption (SE) scheme against chosen ciphertext attacks (**IND-SE-CCA**) is demonstrated as a game between an adversary $\mathcal{A}$ and a challenger $\mathcal{C}$. The game is as follows:

- ***Setup***: *$\mathcal{C}$ takes a security parameter κ and runs the Setup (κ) algorithm to generate the system parameters Params. It provides Params to $\mathcal{A}$. $\mathcal{C}$ then runs KeyGen (U, $Params$) to generate a private and public key pair SK, PK of user U and provides PK to $\mathcal{A}$. SK is kept by $\mathcal{A}$.*
- ***Phase-1***: *$\mathcal{A}$ can adaptively issue queries to the following oracles*:
  - ***Self-encrypt*** *(m, t_w)* ***Oracle***: *$\mathcal{C}$ runs the* ***Self-encrypt*** *($m, t_w, SK, PK, Params$) algorithm to generate ciphertext C and returns it to $\mathcal{A}$.*
  - ***Self-decrypt*** *(C, PK)* ***Oracle***: *$\mathcal{C}$ runs the* ***Self-decrypt*** *(C, SK, PK, $Params$) and returns the output to $\mathcal{A}$.*
- ***Challenge***: *After getting sufficient training, $\mathcal{A}$ submits two messages m_0, m_1 from $\mathcal{M}$ of equal length and a tag $t_w{}^*$ to $\mathcal{C}$. $\mathcal{C}$ picks a random bit $\delta \in \{0, 1\}$ and outputs the ciphertext $C^* =$* ***Self-encrypt*** *($m_\delta, t_w{}^*, SK, PK$).*
- ***Phase-2***: *On receiving the challenge C^*, $\mathcal{A}$ is allowed to access the various oracles provided in* ***Phase-1*** *with the restrictions given below*:
  1. ***Self-decrypt*** *(C^*) query is not allowed.*
  2. ***Self-encrypt*** *(m_δ, $t_w{}^*$) query is not allowed.*
- ***Guess***: *$\mathcal{A}$ outputs its guess δ' and wins the game if $\delta = \delta'$.*

2.6 Security Model for Proxy Re-encryption with Self-encryption

In this section we provide the security model for the SE-PRE scheme. The model involves the security of original ciphertext as well as transformed ciphertext. The ciphertext that can be re-encrypted is called the original ciphertext and the output of the re-encryption is called the transformed ciphertext.

Security of Original Ciphertext The security of Proxy Re-encryption with Self-encryption (SE-PRE) schemes against chosen ciphertext attacks (**IND-SE-PRE-CCA$_O$**) for the original ciphertext is modelled as a game between an adversary $\mathcal{A}$ and a challenger $\mathcal{C}$. The security game is described below:

- ***Setup***: *$\mathcal{C}$ takes a security parameter κ and runs the Setup (κ) algorithm to generate the system parameters Params. The Params is then given to $\mathcal{A}$.*
- ***Phase-1***: *On receiving the system parameters, a target public key PK_T and tag $t_w{}^*$, $\mathcal{A}$ is allowed to access* ***Keygen, Self-encrypt, Self-decrypt, Rekey, Re-encrypt, Re-decrypt*** *algorithms. $\mathcal{A}$ simulates the algorithms as oracles and $\mathcal{A}$ can adaptively issue queries to these oracles. The various oracles provided by $\mathcal{C}$ are:*
  - ***Corrupted KeyGen*** *(U_i): $\mathcal{C}$ runs the KeyGen (U_i, $Params$) to obtain the public and private key pair (PK_i, SK_i). $\mathcal{C}$ returns SK_i and PK_i.*
  - ***Uncorrupted KeyGen*** *(U_i): $\mathcal{C}$ runs the* ***KeyGen*** *(U_i, $Params$) to obtain the public and private key pair (PK_i, SK_i) and returns PK_i to $\mathcal{A}$. SK_i is not provided to $\mathcal{A}$.*
  - ***ReKeyGen*** *(U_i, U_j): $\mathcal{C}$ runs the* ***ReKeyGen*** *(SK_i, PK_i, PK_j, c_w, $Params$) to obtain the re-encryption key $RK_{i\rightarrow j}$ and returns it to $\mathcal{A}$.*
  - ***Self-encrypt*** *(m, t_w, PK_i): $\mathcal{C}$ runs the Self-encrypt (m, t_w, SK_i, PK_i, $Params$) to obtain the ciphertext C and returns it to $\mathcal{A}$.*
  - ***Re-encrypt*** *(C, PK_i, PK_j, c_w): $\mathcal{C}$ runs the* ***Re-encrypt*** *(C, PK_i, c_w, $RK_{i\rightarrow j}$, $Params$) to obtain the ciphertext D and returns it to $\mathcal{A}$. Here, $RK_{i\rightarrow j}$ is the re-encryption key from PK_i to PK_j under the condition c_w.*
  - ***Self-decrypt*** *(C, PK_i): $\mathcal{C}$ runs the* ***Self-decrypt*** *(C, SK_i, PK_i, $Params$) and returns the output to $\mathcal{A}$.*
  - ***Re-decrypt*** *(D, PK_j): $\mathcal{C}$ runs the* ***Re-decrypt*** *(D, SK_j, PK_j, $Params$) and returns the output to $\mathcal{A}$.*
    For the ***ReKey, Encrypt, Re-encrypt, Decrypt, Re-decrypt*** *oracle queries it is required that public keys PK_i and PK_j are generated beforehand.*
- ***Challenge***: *On getting sufficient training, $\mathcal{A}$ will output two equal-length plaintexts $m_0, m_1 \in \mathcal{M}$. Here, the constraint is: PK_T is generated using* ***Uncorrupted Keygen*** *and* ***Rekey*** *(PK_T, PK_j, c_w), is not queried in* ***Phase-1*** *for $c_w = t_w{}^*$ $\mathcal{C}$ flips a random coin $\delta \in \{0, 1\}$, and sets the challenge ciphertext C^* =* **Self-encrypt** *(m_δ, $t_w{}^*$, SK_T, PK_T, $Params$). $\mathcal{C}$ then provide C^* as challenge to $\mathcal{A}$.*
- ***Phase-2***: *$\mathcal{A}$ can adaptively query as in $Phase - 1$ with the following restrictions:*
  1. *$\mathcal{A}$ cannot issue* ***Corrupted KeyGen*** *(U_T) query.*
  2. *$\mathcal{A}$ cannot issue* ***Self-decrypt*** *(C^*, PK_T, $t_w{}^*$) query.*
  3. *$\mathcal{A}$ cannot issue* ***Re-encrypt*** *(C^*, PK_T, PK_j) query on C^* from PK_T to PK_j if PK_j is $Corrupted$.*
  4. *$\mathcal{A}$ cannot issue* ***ReKey*** *(PK_T, PK_j, c_w) query if $c_w = t_w{}^*$.*
  5. *$\mathcal{A}$ cannot issue* ***Re-decrypt*** *query on D^*, PK_j if D^* is the output of* ***Re-encrypt*** *(C^*, PK_T, PK_j, c_w) and $c_w = t_w{}^*$.*
- ***Guess***: *Finally, $\mathcal{A}$ outputs a guess $\delta' \in \{0, 1\}$ and wins if $\delta' = \delta$.*

Security of Transformed Ciphertext The security of transformed of Proxy Re-encryption with Self-encryption (SE-PRE) scheme against chosen ciphertext attacks (**IND-SE-PRE-CCA**$_T$) is modelled as a game between an adversary $\mathcal{A}$ and a challenger $\mathcal{C}$. This is achieved by:

- ***Setup***: *$\mathcal{C}$ takes a security parameter κ and runs the Setup (κ) algorithm and gives the resulting system parameters $Params$, a target public key PK_T and tag $t_w{}^*$ to $\mathcal{A}$.*
  Phase-1: *This phase is similar to the **Phase-1** of **IND-SE-PRE-CCA**$_O$. We do not provide **Re-encrypt** oracle as we are providing all the re-encryption keys for the adversary.*
- ***Challenge***: *Once $\mathcal{A}$ decides $Phase - 1$ is over, it outputs two equal-length plaintexts m_0, $m_1 \in \mathcal{M}$. $\mathcal{C}$ flips a random coin $\delta \in \{0, 1\}$, and sets the challenge ciphertext as follows:*
  - *Compute C^* = **Self-encrypt** ($m_\delta, t_w{}^*, SK_i, PK_i, Params$), (PK_i, SK_i) be the public, private key pair of user U_i and U_i can be honest or corrupt.*
  - *Sets D^* = **Re-encrypt** (C^*, $RK_{i \to T}$) which is then sent to $\mathcal{A}$.*
- ***Phase-2***: *$\mathcal{A}$ adaptively issues queries as in $Phase - 1$, and $\mathcal{C}$ answers them as before with the following restrictions:*
  1. *$\mathcal{A}$ cannot issue **Corrupted KeyGen** (U_T) query.*
  2. *$\mathcal{A}$ cannot issue **Re-decrypt** (D^*, PK_T) query.*
- ***Guess***: *Finally, $\mathcal{A}$ outputs a guess $\delta' \in \{0, 1\}$ and wins the game if $\delta' = \delta$.*

3 The Self-encrypt (SE) Scheme

Self-encrypt scheme is a special kind of encryption primitive that allows a user to store file securely in cloud or any distributed storage. In this approach the owner of the file uses his/her private key to encrypt the file. This significantly reduces the computation involved in storing the file. We provide the self encryption scheme and the prove its CCA security in the random oracle model.

3.1 The Scheme

The SE scheme consist of the following algorithms:

- **Setup** (κ):
  - Let $\mathbb{G}$ be an additive cyclic group of prime order q. Let P be a generator of group $\mathbb{G}$.

- Let Δ = ⟨**Sym. Encrypt**, **Sym. Decrypt**⟩ be any symmetric key encryption scheme. We may assume that Δ is a symmetric key encryption algorithm that uses messages of block size k.
- Choose the hash functions,

$$H_1 : \{0,1\}^{l_t} \times \mathbb{Z}_q{}^* \rightarrow \mathbb{Z}_q{}^*$$
$$H_2 : \mathbb{Z}_q{}^* \rightarrow \{0,1\}^{l_k}$$
$$H_3 : \{0,1\}^{l_m} \times \mathbb{G} \rightarrow \{0,1\}^{l_3}$$

- Here l_t is the size of the tag, l_m is the size of the message and l_k is the size of the symmetric key used by the symmetric key encryption scheme Δ. Also, l_3 is dependent on the security parameter κ.
- Output $Params = \langle q, \mathbb{G}, P, H_1(), H_2(), H_3(), \Delta\rangle$

- **KeyGen** $(U, Params)$: The $KeyGen$ algorithm generates the public and private key of the user U by performing the following steps:
  - Choose a random integer $x \stackrel{R}{\leftarrow} \mathbb{Z}_q{}^*$
  - Output $PK = \langle X = xP\rangle$ and $SK = \langle x\rangle$.
- **Self-encrypt** $(m, t_w, SK, PK, Params)$: On input of message m, tag t_w, private key $SK = x$ of user U, public key $PK = xP$ of user U and the public parameters $Params$ this algorithm will generate the self encryption as follows:
  - Choose random $t \in \mathbb{Z}_q{}^*$
  - Set $h_t = H_1(t_w, x)$.
  - Set $C_1 = t + h_t$.
  - Compute $Key = H_2(t)$
  - $C_2 = \{\hat{C}_i\}_{(for\, i=1\, to\, l)}$ and $\hat{C}_i$ = Sym.Encrypt (M_i, Key) for all $i = 1$ to l, l is the number of blocks. Assume that $m = M_1, M_2, \ldots, M_l$ where $|M_i| = k$ and k is the block size of Δ.
  - $C_3 = H_3(m, t)$
  - Output the ciphertext $C = \langle C_1, C_2, C_3, t_w\rangle$.
- **Self-decrypt** $(C, SK_i, Params)$: Self decryption algorithm is used to decrypt the files that are previously encrypted by the user U using his/her private key. This algorithm does the following:
  - $h_t = H_1(t_w, SK_i)$.
  - $t = C_1 - h_t$
  - $Key = H_2(t)$
  - Compute M_i = **Sym.Decrypt** $(\hat{C}_i, Key)$ for all i = 1 to l and construct $m = M_1, M_2, \ldots, M_l$.
  - If $C_3 \stackrel{?}{=} H_3(m, t)$ then, output m. Else, Output $\perp$.

Correctness of t:

$$\begin{aligned} RHS &= C_1 - h_t \\ &= (t + h_t) - h_t \\ &= t; \\ &= LHS \end{aligned}$$

3.2 *Security Proof*

Theorem 1 *If there exists a* (γ, ϵ) *adversary* $\mathcal{A}$ *with an advantage* ϵ *that can break the* ***IND-SE-CCA*** *security of the SE scheme, then* $\mathcal{C}$ *can solve the discrete log problem with advantage* ϵ' *where,*

$$\epsilon' \geq \epsilon$$

Proof In this section we formally prove the security of **SE** scheme in the random oracle model. The **IND-SE-CCA** security of the SE scheme is reduced to the discrete logarithm problem(DLP). The challenger $\mathcal{A}$ is given with the instance of DLP (i.e given (q, P, Y) such that q is a large prime, $P, Y \in \mathbb{G}$, find a such that $Y = aP$.) If there exist an adversary $\mathcal{A}$ that can break the $IND-SE-CCA$ security of the **SE** scheme, then $\mathcal{C}$ can make use of $\mathcal{A}$ to solve the discrete logarithm problem, which is assumed to be hard. Thus the existence of such adversary is not possible.

The challenger $\mathcal{C}$ sets the public key $PK = Y$(PK = aP) and the corresponding private key $SK = x = a$ (which is not known to $\mathcal{C}$). $\mathcal{C}$ then provides PK to $\mathcal{A}$. $\mathcal{A}$ has access to various algorithms of **SE** and the hash functions as oracles. $\mathcal{C}$ simulates the hash functions and the **Self-encrypt, Self-decrypt** algorithms as described below:

- **Phase-1**: $\mathcal{A}$ is given to access all the oracles as defined in the security model **IND-SE-CCA**. Here it should be noted that $\mathcal{C}$ which does not have the knowledge of private key $SK = a$ provides the functionalities **Self-encrypt, Self-decrypt** algorithm.
  - The hash functions involved in the **SE** scheme are simulated as random oracles. To provide consistent output, $\mathcal{C}$ maintains the lists L_{H_1}, L_{H_2} and L_{H_3} corresponding to the hash function H_1, H_2 and H_3 involved in the **SE** scheme.
    - * H_1 Oracle: When a query with input (t_w, x) is made, the tuple $\langle t_w, x, h_t \rangle$ is retrieved from L_{H_1} and h_t is returned, if (t_w, x) is already there in L_{H_1} list. Otherwise, $\mathcal{C}$ does the following:
      - · If $xP = Y$, then abort. This is because $\mathcal{C}$ obtains the solution to DLP i.e $x = a$.
      - · Pick $h_t \in \mathbb{G}$.
      - · If h_t is already present in L_{H_1} list, go to previous step.
      - · Store $\langle t_w, x, h_t \rangle$ in L_{H_1} list and output h_t.

∗ H_2 Oracle: When a query with t is made, $\mathcal{C}$ the tuple $\langle t,\ Key\rangle$ from LH_2 list is retrieved and will return Key, if (t) is already present in L_{H_2} list. Otherwise, $\mathcal{C}$ does the following:

· Pick $Key \in \{0, 1\}^{l_k}$.

· If Key is already present in L_{H_2} list, go to previous step.

· Store $\langle t,\ Key\rangle$ in L_{H_2} list and return Key.

∗ H_3 Oracle: When a query with input (m, T) is made, $\mathcal{C}$ retrieves the tuple $\langle m,\ T,\ \alpha\rangle$ from L_{H_3} list and returns α, if (m, T) is already present in L_{H_3} list. Otherwise, $\mathcal{C}$ does the following:

· Pick $\alpha \in \{0, 1\}^{l_3}$.

· If α is already present in L_{H_3} list, go to previous step.

· Store $\langle m,\ T,\ \alpha\rangle$ in L_{H_3} list and return α.

– **Self-encrypt** Oracle: When a **Self-encrypt** query is made with (m, t_w) as input, $\mathcal{C}$ does the following:

∗ Choose random $t \in \mathbb{Z}_q{}^*$.

∗ Set $h_t = H_1(t_w,\ x)$.

∗ Set $C_1 = t + h_t$.

∗ Compute $Key = H_2(t)$.

∗ $C_2 = \{\hat{C}_i\}_{(for\ i=1\ to\ l)}$ and $\hat{C}_i$ = Sym.Encrypt $(M_i,\ Key)$ for all $i = 1$ to l, l is the number of blocks. Assume that $m = M_1, M_2, \ldots, M_l$ where $|M_i| = k$ and k is the block size of Δ.

∗ $C_3 = H_3(m,\ t)$.

∗ Output the ciphertext $C = \langle C_1,\ C_2,\ .C_3,\ t_w\rangle$.

∗ Output the self-encrypted ciphertext C to $\mathcal{A}$.

– **Self-decrypt** Oracle: When a **Self-decrypt** query is made with $C = \langle C_1,\ C_2,\ C_3,\ t_w\rangle$ as input, $\mathcal{C}$ performs the following:

∗ If C is in $L_{Encrypt}$ list, pick m corresponding to C from the tuple $\langle C,\ m\rangle$ in $L_{Encrypt}$ list and output m.

∗ If $(t_w,\ -)$ is present in L_{H_1} list then, retrieve h_t corresponding to $(t_w,\ -)$ from L_{H_1} list. Else, it returns $\perp$.

∗ $T = C_1 - ht$.

∗ $Key = H_2(t)$.

∗ Compute M_i = **Sym.Decrypt** $(\hat{C}_i,\ Key)$ for all i=1 to l and construct $m = M_1\ M_2 \ldots M_l$.

∗ If $C_3 \stackrel{?}{=} H_3(m,\ t)$ then, output m. Else, it output $\perp$.

- **Challenge Phase**: After the first phase of training is over, $\mathcal{A}$ provides $m_0,\ m_1 \in \mathcal{M},\ t_w{}^*$ such that $(m_0, t_w{}^*)$ or $(m_1, t_w{}^*)$ was not queried to **Self-encrypt** oracle during **Phase-1** and provides to $\mathcal{C}$. $\mathcal{C}$ now generates the challenge ciphertext C^* = **Self-encrypt** $(m_\delta, t_w{}^*)$ and $\delta\ \in_R\ \{0, 1\}$
- **Phase-2**: $\mathcal{A}$ can interact with all the oracles as in **Phase-1** but with the following restrictions:

  – $\mathcal{A}$ cannot make the query **Self-decrypt** (C^*)
  – $\mathcal{A}$ cannot make the query **Self-encrypt** $(m_\delta, t_w{}^*)$, $\delta\ \in\ \{0, 1\}$

- **Guess**: Once **Phase-2** is over, $\mathcal{A}$ output its guess δ'. $\mathcal{A}$ wins the game if $\delta = \delta'$.

4 The Proxy Re-encryption with Self Encryption Scheme (SE-PRE)

In this section we present a proxy re-encryption scheme which uses the self encryption proposed in Sect. 3. The **SE** scheme is modified such a way that it allows verifiability of ciphertext by proxy during re-encryption without knowing the message. It helps in achieving CCA security of SE-PRE. This also helps in avoiding the DDOS attack being launched on Proxy's service. The proxy is equipped with a method to identify invalid ciphertext so that it will serve its functionality only to valid input. Also.the **SE-PRE** algorithm can be deployed in a simple and efficient manner than using the traditional PRE schemes available till date.

4.1 The Scheme

In this section we present the proxy re-encryption scheme **SE-PRE** that uses private encryption algorithm. The **SE-PRE** proposed here uses a novel approach, consisting of the following algorithms.

- **Setup** (κ):
  - Let $\mathbb{G}$ be an additive cyclic group of prime order q. Let P be a generator of group $\mathbb{G}$.
  - Let Δ = ⟨**Sym. Encrypt**, **Sym. Decrypt**⟩ be any symmetric key encryption scheme. We may assume that Δ is a symmetric encryption algorithm working on block of size k.
  - Choose the following hash functions:

$$
\begin{aligned}
&H_0 : \{0, 1\}^{l_t} \rightarrow \{0, 1\}^{l_0},\\
&H_1 : \{0, 1\}^{l_t} \times \mathbb{Z}_q{}^* \times \mathbb{G} \rightarrow \mathbb{Z}_q{}^*,\\
&H_2 : \mathbb{Z}_q{}^* \rightarrow \{0, 1\}^{l_k},\\
&H_3 : \{0, 1\}^{l_m} \times \mathbb{Z}_q{}^* \rightarrow \{0, 1\}^{l_3},\\
&H_4 : \mathbb{Z}_q{}^* \times \{0, 1\}^{(l_c+l_3+l_5)} \times \mathbb{G} \rightarrow \mathbb{Z}_q{}^*,\\
&H_5 : \{0, 1\}^{l_t} \times \mathbb{Z}_q{}^* \times \mathbb{G} \rightarrow \{0, 1\}^{l_5},\\
&H_6 : \{0, 1\}^{l_w} \times \mathbb{G} \times \mathbb{G} \times \mathbb{G} \rightarrow \mathbb{Z}_q{}^*,\\
&H_7 : \mathbb{Z}_q{}^* \times \mathbb{G} \rightarrow \mathbb{Z}_q{}^*,\\
&H_8 : \mathbb{G} \times \mathbb{G} \rightarrow \{0, 1\}^{(l_w+l_p)},\\
&H_9 : \{0, 1\}^{l_u} \times \mathbb{Z}_q{}^* \rightarrow \mathbb{Z}_q{}^*,\\
&H_c : \{0, 1\}^* \rightarrow \{0, 1\}^{l_c}.
\end{aligned}
$$

- Here l_t is size of the tag, l_m is size of the message, l_c is the size of the ciphertext, l_p is κ and l_k is size of the symmetric key used in the encryption scheme Δ. Also, l_ω, l_u, l_0, l_3 and l_5 are dependent on the security parameter κ.
- Output $Params = \langle q, P, G, P, H_i()_{(for\, i=0\, to\, 9)}, H_c(), \Delta\rangle$

- **KeyGen** $(U_i, Params)$: The $KeyGen$ algorithm generates the public and private key of the user U_i by performing the following:
  - Choose a random integer $x_i \xleftarrow{R} \mathbb{Z}_q{}^*$
  - Output $PK_i = \langle X_i = x_i P\rangle$ and $SK = \langle x_i\rangle$.
- **RekeyGen** $(SK_i, PK_i, PK_j, c_w, Params)$: This algorithm generates the re-encryption key required to translate a ciphertext of user U_i into a ciphertext of user U_j. This is run by the user U_i. The ciphertext to be re-encrypted is encrypted under the public key PK_i of user U_i and with the condition c_w, which are specified by user U_i. This algorithm works as follows:
  - Choose $\omega \xleftarrow{R} \in \{0, 1\}^{l_\omega}$
  - Compute $h_c = H_1(c_w, x_i, X_i) \in \mathbb{Z}_q*$
  - Compute $r = H_6(\omega, x_i X_j, X_i, X_j) \in \mathbb{Z}_q*$
  - Compute $s = H_7(r, X_j) \in \mathbb{Z}_q*$
  - Compute $\gamma = rX_j$
  - Compute the re-encryption key $RK_{i\rightarrow j} = \langle R_1, R_2, R_3, R_4, R_5, R_6\rangle$ where,

$$
\begin{aligned}
R_1 &= s - h_c \in \mathbb{Z}_q* \\
R_2 &= rP \in \mathbb{G} \\
R_3 &= (\omega || X_i) \oplus H_8(\gamma, X_j) \in \{0, 1\}^{l_\omega + l_g} \\
R_4 &= H_6(\omega, \gamma, X_i, X_j) \in \mathbb{Z}_q* \\
R_5 &= H_5(t_w, x_i, X_i) \in \{0, 1\}^{l_5} \\
R_6 &= H_0(t_w)
\end{aligned}
$$

  - Output the re-encryption key $RK_{i\rightarrow j} = \langle R_1, R_2, R_3, R_4, R_5, R_6\rangle$
- **Self-encrypt** $(m, t_w, SK_i, PK_i, Params)$: On input of message m, tag t_w, private key SK_i, public key PK_i of user U_i and the public parameters $Params$
  - Choose random $\omega \in \mathbb{Z}_q*$
  - Set $h_t = H_1(t_w, x_i, X_i) \in \mathbb{Z}_q*$
  - Compute $C_1 = t + h_t$.
  - Compute $Key = H_2(t)$
  - Compute $C_2 = \{\hat{C}_i\}_{(for\, i=1\, to\, l)}$ and $\hat{C}_i$ = **Sym.Encrypt** (M_i, Key) for all $i = 1$ to l, l is the number of blocks. Assume that $m = M_1, M_2, \ldots, M_l$ where $|M_i| = k$ and k is the block size of Δ.
  - Set $C_3 = H_3(m, t)$
  - Find $\alpha = H_5(t_w, x_i, X_i) \in \{0, 1\}^{l_5}$

- $C_4 = H_4(C_1, C_2, C_3, \alpha, X)$
- Set $C_5 = H_0(t_w)$
- Output the ciphertext $C = \langle C_1, H_c(C_2), C_3, C_4, C_5 \rangle$.

- **Re-encrypt** $(C, PK_i, PK_j, c_w, RK_{i\rightarrow j}, Params)$: This algorithm is run by the proxy which is given with the re-encryption key $RK_{i\rightarrow j}$ by user U_i. This generates the re-encryption of a ciphertext encrypted under public key PK_i of user U_i under the condition c_w into a ciphertext encrypted under public key PK_j of user U_j. This algorithm does not perform any complex computation and this greatly reduces the computational overhead on the entity that performs the role of a proxy. This algorithm does the following computations :
  - If $C_4 \neq H_4(C_1, H_c(C_2), C_3, R_5, t_w, X)$ OR $C_5 \neq R_6$, then it returns $\perp$
  - Set $D_2 = C_2$, $D_3 = C_3$, $D_4 = R_2$, $D_5 = R_3$
  - Choose $u \in \{0, 1\}^{l_u}$
  - Compute $\beta = H_9(u, R_4) \in \mathbb{Z}_q{}^*$
  - Compute $D_1 = \beta(C_1 + R_1) \in \mathbb{Z}_q{}^*$
  - Set $D_6 = u$
  - Output the re-encrypted ciphertext $D = \langle D_1, D_2, D_3, D_4, D_5, D_6 \rangle$

- **Self-decrypt** $(C, SK_i, Params)$: Self-decrypt algorithm is used to decrypt the self-encrypted ciphertext C of a user that is stored by him in the cloud. This algorithm performs the following:
  - Find $\alpha = H_5(t_w, x_i, X_i) \in \{0, 1\}^{l_5}$
  - If $C_4 \neq H_4(C_1, C_2, C_3, \alpha, t_w, X)$, then it returns $\perp$
  - $h_t = H_1(t_w, SK_i)$.
  - $t = C_1 - h_t$
  - $Key = H_2(t)$
  - Compute M_i = **Sym.Decrypt** $(\hat{C}_i, Key)$ for all i=1 to l and construct $m = M_1 M_2 \ldots M_l$.
  - If $C_3 \stackrel{?}{=} H_3(m, t)$ then, output m. Else, Output $\perp$.
    <u>**Correctness of**</u> t:

$$\begin{aligned} RHS &= C_1 - h_t \\ &= (t + h_t) - h_t \\ &= t \\ &= LHS \end{aligned}$$

  - **Re-decrypt** $(D, SK_j, Params)$: The **Re-decrypt** algorithm is used to decrypt the re-encrypted ciphertext D. This algorithm does the following:
    · Compute $\gamma = x_j D_4$
    · Compute $\omega || X_i = D_5 \oplus H_8(\gamma, X_j)$
    · Compute $r = H_6(\omega, x_j X_i, X_i, X_j) \in \mathbb{Z}_q*$
    · Compute $s = H_7(r, X_j) \in \mathbb{Z}_q*$
    · $\rho = H_6(\omega, \gamma, X_i, X_j) \in \mathbb{Z}_q*$

- Compute $\beta = H_9(D_6, \rho) \in \mathbb{Z}_q*$
- Compute $t = \beta^{-1}(D_1) - s$
- Find $Key = H_2(t)$
- Compute M_i = **Sym.Decrypt** $(\hat{C}_i, Key)$ for all i = 1 to l and construct $m = M_1, M_2, \ldots, M_l$.
- If $(C_3 \stackrel{?}{=} H_3(m, t))$ then, output m. Else, it returns $\perp$.

Correctness of T:

$$\begin{aligned} RHS &= \beta^{-1} D_1 - s \\ &= \beta^{-1}[\beta(C_1 + R_1)] - s \\ &= [(t + h_t) + (s - h_c)] - s;\ Here\, h_t = h_c \\ &= (t + s) - s \\ &= t \\ &= LHS \end{aligned}$$

4.2 Security Proof

Security of the Original Ciphertext

Theorem 2 *If a* (γ, ϵ) *adversary* $\mathcal{A}$ *with an advantage* ϵ *breaks the* ***IND-SE-PRE-CCA$_O$*** *security of the SE-PRE scheme in time* γ, *then* $\mathcal{C}$ *can solve the discrete log problem or CDH with advantage* ϵ' *where,*

$$\epsilon' \geq \frac{1}{q_t}\epsilon$$

Here, q_t *is the number of queries to* H_6 *oracle.*

Proof Due to space constraints, the proof of the theorem is given in the full version of the paper.

Security of the Transformed Ciphertext

Theorem 3 *If a* (t, ϵ) *adversary* $\mathcal{A}$ *with an advantage* ϵ *breaks the* ***IND-SE-PRE-CCA$_T$*** *security of the SE-PRE scheme, then* $\mathcal{C}$ *can solve the Computational Diffie Hellman(CDH) problem with advantage* ϵ' *where,*

$$\epsilon' \geq \frac{1}{q_t}\epsilon$$

Here, q_t *is the number of queries to* H_6 *oracle.*

Proof Due to space constraints, the proof of the theorem is given in the full version of the paper.

5 Experimental Analysis

In this section we provide the implementation results and time taken by various algorithms in **SE** and **SE-PRE** scheme. We compare the efficiency of our CCA secure **SE** scheme with the traditional CPA secure El-Gamal scheme (Weaker security than CCA) and report the same in Table 1. Also, we have compared our **SE-PRE** scheme with the only non-pairing unidirectional CCA secure PRE scheme by Selvi et al. [3] available. This is reported in Table 2. It is a known fact that pairing is very expensive than other group operations and hence we are not taking any pairing based schemes into consideration. The implementations are done on 2.4 GHz Intel Core i7 quad-core processor and the results have been reported below. The programming language used is GO language [10], and the programming tool is Goland 2018.2. The cryptographic protocols are implemented using the edwards25519-curve [11], which is the current standard deployed in cryptocurrencies [12] for fast performances. From the performance comparison in Table 1, we note that our CCA secure self-encryption **SE** scheme is more efficient than the existing CPA-secure El-Gamal encryption scheme [13]. Also, from Table 2, it is evident that our self-proxy re-encryption **SE-PRE** scheme without bilinear pairing is more efficient than the existing pairing-free PRE scheme by Selvi et al. [3]. From the shown results, it is evident that our **SE** encryption scheme is practical and suitable for cloud based scenarios where the user themselves store their files. Also, the **SE-PRE** scheme provides a very efficient approach to share encrypted files mainly in block-chain.

Table 1 Performance evaluation of the CPA secure El-Gamal encryption scheme and our self-encryption scheme (all timings reported are in microseconds)

Algorithm	CPA-secure El-Gamal scheme	Our CCA secure **SE** scheme
Key generation	612.947	591.677
Encryption	420.307	65.416
Decryption	300.052	41.65

Table 2 Performance evaluation of the efficient pairing-free unidirectional PRE scheme due to Chow et al. and our scheme (all timings reported are in microseconds)

Algorithm	CCA-secure Selvi et al. scheme	Our CCA secure **SE-PRE** scheme
Key generation	714.271	579.702
First level encryption	1044.695	87.85
First level decryption	1554.78	60.356
Re-encryption key generation	478.368	796.036
Re-encryption	1087.52	23.216
Re-decryption	1077.05	745.031

6 Conclusion

In this paper, we have given a self encryption scheme **SE** based on discrete logarithm (DLP) assumption and then extended it to a Proxy Re-encryption(**SE-PRE**) scheme suitable for block chain and distributed storage. First, we formally prove the CCA security of the **SE** and then the security of **SE-PRE** scheme in the random oracle model. We have also implemented our **SE-PRE** scheme using GO language. From the results of our implementation, it is evident that our **SE-PRE** scheme is much efficient than the techniques available in literature till date. This makes it more suitable for distributed applications. It will be interesting to see how one can design a multi-recipient or broadcast PRE based on the self encryption approach that will provide high efficiency gain in decentralised platforms.

Acknowledgements We would like to thank the anonymous reviewers whose comments helped us improve this paper.

References

1. 0box application by 0chain: https://0chain.net/zerobox
2. Ochain website: https://0chain.net
3. Selvi, S.S.D., Paul, A., Pandu Rangan, C.: A provably-secure unidirectional proxy re-encryption scheme without pairing in the random oracle model. In: CANS, Lecture Notes in Computer Science, vol. 11261, pp. 459–469. Springer (2017)
4. Blaze, M., Bleumer, G., Strauss, M.: Divertible protocols and atomic proxy cryptography. In: Advances in Cryptology—EUROCRYPT'98, International Conference on the Theory and Application of Cryptographic Techniques, Espoo, Finland, May 31 - June 4, 1998, Proceeding, pp. 127–144 (1998)
5. Ateniese, G., Fu, K., Green, M., Hohenberger, S.: Improved proxy re-encryption schemes with applications to secure distributed storage. In: Proceedings of the Network and Distributed System Security Symposium, p. 2005. NDSS, San Diego, California, USA (2005)
6. Ateniese, G., Fu, K., Green, M., Hohenberger, S.: Improved proxy re-encryption schemes with applications to secure distributed storage. ACM Trans. Inf. Syst. Secur. **9**(1), 1–30 (2006)
7. Selvi, S.S.D., Paul, A., Pandu Rangan, C.: An efficient non-transferable proxy re-encryption scheme. In: Proceedings of the 8th International Conference on Applications and Techniques in Information Security—ATIS 2017, Auckland, New Zealand, July 6-7, 2017, pp. 35–47 (2017)
8. Nuñez, D., Agudo, I., López, J.: Proxy re-encryption: Analysis of constructions and its application to secure access delegation. J. Netw. Comput. Appl. **87**, 193–209 (2017)
9. IAgudo, I., Nuez, D., Lopez, J.: A parametric family of attack models for proxy re-encryption. Cryptology ePrint Archive, Report 2016/293 (2016). https://eprint.iacr.org/2016/293
10. Bernstein, D.J.: The GO programming language. https://golang.org/
11. Bernstein, D.J.: Curve25519: new diffie-hellman speed records. In: Proceedings of the 9th International Conference on Theory and Practice of Public-Key Cryptography—PKC 2006, New York, NY, USA, April 24-26, 2006, pp. 207–228 (2006)
12. Mayer, H.: Ecdsa security in bitcoin and ethereum: a research survey. *CoinFaabrik*, June 28 (2016)
13. El Gamal, T.: A public key cryptosystem and a signature scheme based on discrete logarithms. IEEE Trans. Inf. Theory **31**(4), 469–472 (1985)

Digital Currencies: A Multivariate GARCH Approach

Stamatis Papangelou and Sofia Papadaki

Abstract In this paper we will present quantifiable linkages between five different cryptocurrencies, those being Bitcoin, Ethereum, Ripple, Dash and Monero. Initially, we conduct a review of the existing related work. As the concept of cryptocurrencies is fairly new, the relevant literature is very restricted. Attempting to bridge a gap in the existing methodologies, we extract our results by using a five-variable conditional asymmetric GARCH-CCC model, and we conclude that a strong influence exists, of the individual past shocks and volatility in all digital currencies that we include in the research. As estimated by the conditional time-varying covariance, we observe that the interlinkages between the cryptocurrencies are very strong and all covariances follow similar patterns resulting in a highly interdependent and volatile system of assets that is not suitable for a diversified portfolio.

Keywords Volatility · Multivariate GARCH model · Cryptocurrencies · Bitcoin

1 Introduction

In most parts of the planet, cryptocurrencies are in the spotlight of various financial and economic news topics. After an extensive research on the literature we wanted to give our own definition on cryptocurrencies. So according to our statement a cryptocurrency can be defined as "a digital asset that can be used as a mean of exchanging value in a digitally encrypted environment so the creation of additional units and transactions can exist in a trustworthy decentralized space".

Everything started with Bitcoin. Bitcoin, that was first released in January 2009, uses a peer-to-peer electronic cash system [26] that employs blockchain as a public ledger to record Bitcoin's transactions. Bitcoin revolutionized the digital currencies,

S. Papangelou (✉)
University of Macedonia, Egnatia 156, 546 36 Thessaloniki, Greece
e-mail: manosp@pm.me

S. Papadaki
National and Kapodistrian University of Athens, Panepistimiou 30, 106 79 Athens, Greece

P. Pardalos et al. (eds.), *Mathematical Research for Blockchain Economy*,
Springer Proceedings in Business and Economics,
https://doi.org/10.1007/978-3-030-37110-4_5

and its price and capitalization have outperformed all other cryptocurrencies that followed. Since cryptocurrencies and Blockchain technology in general is very recently born, research in that area is still very limited, so it is very interesting to see what academics and researchers will come up with next. In this study we will not examine the principles and technologies behind digital currencies. Instead, we are going to focus on their investing and pricing behaviors [2].

In particular, we are going to study individual price fluctuations on a group of some of the most popular cryptocurrencies [17]. More specifically, we are going to study five different digital coins. The first and major one is Bitcoin, that has been mentioned above. The second one is Ethereum. It was introduced on July 2015 as a blockchain-based distributed computing platform and operating system featuring smart contract (scripting) functionality. It supports an updated version of Nakamoto's consensus via transaction-based state transitions [35]. Ether is the cryptocurrency whose blockchain is given by the Ethereum platform and it's a fundamental aspect in the operation of Ethereum [15]. Ripple, that was released in 2012, uses an open-source protocol as basic infrastructure technology for interbank transactions. Ripple is based on a public ledger that uses a consensus process that allows for payments, exchanges and remittance in distributed process. Companies like UniCredit, UBS and Santander adopted Ripple as settlement infrastructure technology. Dash, that was released on January 2014 and whose name came from "Digital Cash", is an open source peer-to-peer cryptocurrency which is mostly focused on the payment industry. Dash offers instant transactions with the service "InstantSend", private transactions with the service "PrivateSend" and they operate a self-governing and self-funding mechanism that fosters the creation of independent entities that serve network [12]. Lastly, Monero, that was created in April 2014, is an open-source cryptocurrency that is mostly focused on privacy and decentralization. Unlike Bitcoin, Monero is based on the "CryptoNight proof-of-work hash algorithm" which has significant algorithmic differences relating to blockchain obfuscation resulting in high levels of privacy [25].

The goal of this paper is to show using a quantifiable method the interactions and bonds connecting this system of cryptocurrencies not taking into account other risk -neutral cryptocurrencies like "stable coins" [19]. Therefore, we are going to create a multivariate system [20], which we will model with a generalized autoregressive conditional heteroskedasticity (GARCH) model and then study the results of those well-known digital currencies [31].

Furthermore, In Sect. 2 we are going to present a brief history of past research done on the subject. Section 3 will discuss the data that we have used, Sect. 4 will analyze the methodology that we are going to use, then In Sect. 5 we will analyze the empirical results and finally, In Sect. 6, we will present our conclusions.

2 Related Work

As we have defined in the previous section, cryptocurrencies are considered to be assets, but most of the electronic cash, including Bitcoin, were initially designed to be a medium of exchange. So, the question arises whether Bitcoin's prices behave more like an asset or a currency. This question was firstly answered from Glaser et al. [18], who approached the subject of the intentions that users have when changing currency into a digital currency. In order to investigate that, they collected trading data from Bitcoin Blockchain, visitor statistics from Bitcoin Wikipedia and dates of important Bitcoin events. Then they researched the link between intra-network Bitcoin transactions and on exchange trading volumes and finally they analyzed whether new users have an impact on both types of volume. The results indicate that the new users tend to trade Bitcoin with speculating intentions and have low interest to rely on the network as means for paying goods and services [18]. A year later Baek and Elbeck [1] did their own research for examining the same question. They approached the problem by comparing Bitcoin's prices with the S&P 500 Index. Their results show that Bitcoin is 26 times more volatile than S&P 500 Index and that Bitcoin returns are internally driven by buyers and sellers and are not influenced by fundamental economic factors [1]. Lastly, further research on the subject was conducted by Dyhrberg [13], who investigated the hedging capabilities of bitcoin, the arbitrage possibilities [17] and the diversification possibilities [27]. The researches mentioned above suggest that a volatility approach similar to that used for assets may be most appropriate.

Furthermore, on the earlier studies we see that a variety of GARCH-type models have been used. More in detail, Chu et al. [9] and Katsiampa [21] have examined the best suited heteroskedasticity model to be fitted for cryptocurrency volatility analysis. Their results suggest that the most optimal models are the AR-CGARCH, IGARCH and GJRGARCH. However, the research that has been conducted so far uses different models. Glaser et al. [18] and Gronwald [19] use a linear GARCH. Threshold GARCH (TGARCH) and Exponential GARCH (EGARCH) was used by Dyhrberg [13], Bouoiyour et al. [7]. Finally, the Component with Multiple Threshold-GARCH (CMT-GARCH) was only used by Bouoiyour et al. [7].

As mentioned before, cryptocurrency and blockchain technology in general is very novel concept [28], and as a result, the relevant literature is so far very sparse. Moreover, almost all of the studies that were described above have conducted their research using different univariate approaches, with the exception of Stavroyiannis and Babalos [30], who have done a multivariate BEKK approach of Bitcoin prices and the S&P 500 Index. Thus, we argue that the above literature misses a multivariate approach inside the cryptocurrency market, so that the strong linkages between the prices and volatility of some of the major digital currencies can be quantified [29].

3 Data

In our study we use five different "widely known" digital currencies so we will able to build a basic portfolio with them [10]. We wanted for the currencies to differ in capitalization and also not to be derivatives of Bitcoin [14]. So, we selected the three digital currencies with different capitalizations and different utility and security designs. Those are Bitcoin, Ethereum and Ripple with USD 141.2bn, USD 69.7bn and USD 27.2bn of capitalizations respectively (data from May 2018) [11]. Furthermore, we selected two additional cryptocurrencies with smaller, in comparison to Bitcoin, capitalization cryptocurrencies. These are Dash and Monero with 3.3 Billion USD and 3.2 Billion USD of capitalizations accordingly (data from May 2018) [11].

For the digital currencies that were described above we collected all chronological daily price data (note: the cryptocurrency market works around the clock so the prices are not closing prices, but daily average) from 8/08/2015 to 4/16/2018 resulting in 983 observations. It is worth mentioning that the price data of cryptocurrencies are significantly different from a regular stock because the digital currency markets trade around the clock seven days a week. The data was taken from the website "Coinmarketcap" and are available at https://coinmarketcap.com/ [11].

Figure 1 presents the time plots of the series of each and every digital currency we include in our research. The first significant observation is that all of our prices evolve in a very similar way, all of them do not have major price changes prior to the first part of 2017, when all series started a dramatic upshift [4]. We can also see that all the prices in our series reached their apex at the end of 2017 just before switching to 2018, and then all of them started a downtrend that continued until the end of the sample series. As a result of the above we expect very strong linkages between the series resulting to highly significant coefficients and high amount of correlation between the series.

Figure 2 displays the returns of the currencies prices in USD. The price returns were calculated by taking the first differences of the natural logarithm of the raw price data. From the first look of the graphs we see that all digital currencies in our research are very volatile, without systemic patterns between them. Furthermore, we see that the most stable among them is Bitcoin which shows the highest values of volatility in the last part of 2017. Dash also has its greatest values of volatility in the last part of 2017 but shows higher values of volatility compared to Bitcoin. Ether's chart shows that the volatility is more evenly distributed across the window of study with the highest value of volatility coming from the middle of 2015. Monero is volatile across the window of study with the highest peaks in the middle of 2016. Lastly Ripple is the most volatile cryptocurrency in our study with the highest values observed from the first part of 2017 onwards.

Table 1 contains the summary statistics for the return prices of our variables. During the period we study, the performances of the digital currencies are measured by the average returns, with Ethereum having the highest values and Bitcoin the lowest ones. Ripple has the biggest value of standard deviation with 7.9%, followed closely by Monero, with a value of 7.4%. Bitcoin is the digital currency that shows

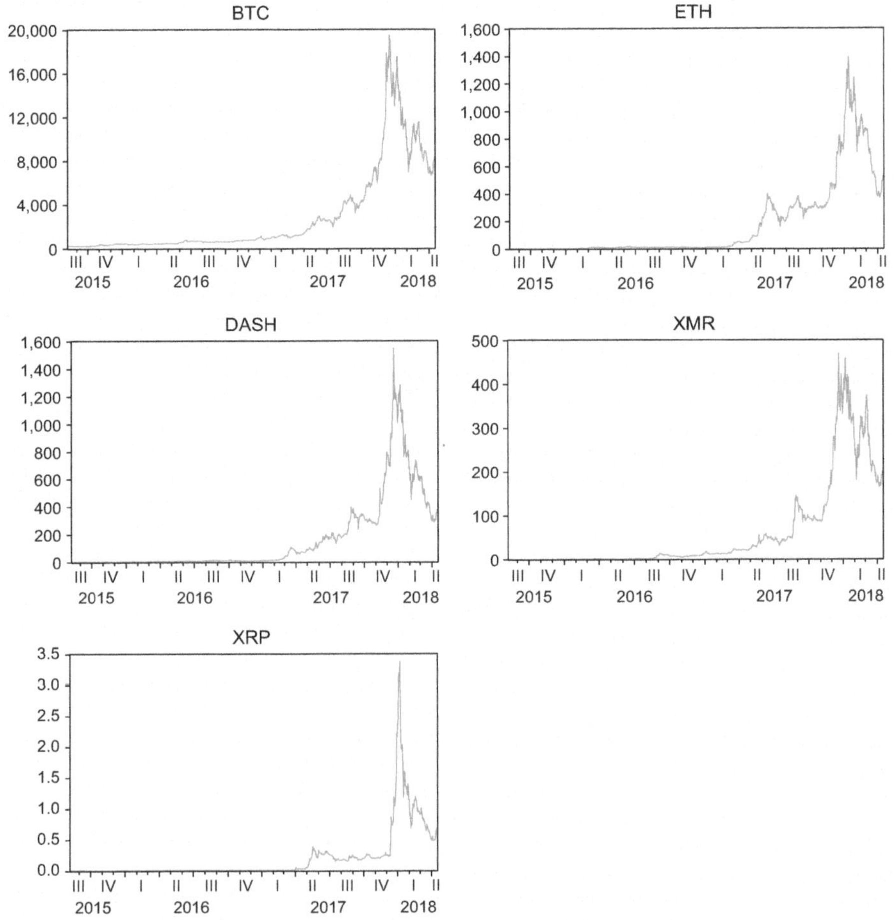

Fig. 1 Time plots of prices from August 2015 to April 2018. *Note* The prices of BTC, ETH, DASH, XMR and XRP correspond to Bitcoin, Ethereum, Dash, Ripple and Monero respectively

the lowest standard deviation value of 4%. In all cases Jarque-Bera statistics reject the null hypothesis of normal distribution of the return prices. Bitcoin has a negative skewness, indicating that is more common to observe large negative shocks in the returns than positive ones. In contrast, the rest of our variables are positively skewed. Moreover, all our variables are leptokurtic, meaning that we observe fatter tails with higher peaks in the distribution, as the kurtosis is greater than 3. The data that were displayed above show that GARCH models will be more suitable. By modelling with GARCH, the non-zero skewness statistics indicate that an ARCH order higher than one in the conditional variance equations is needed [3].

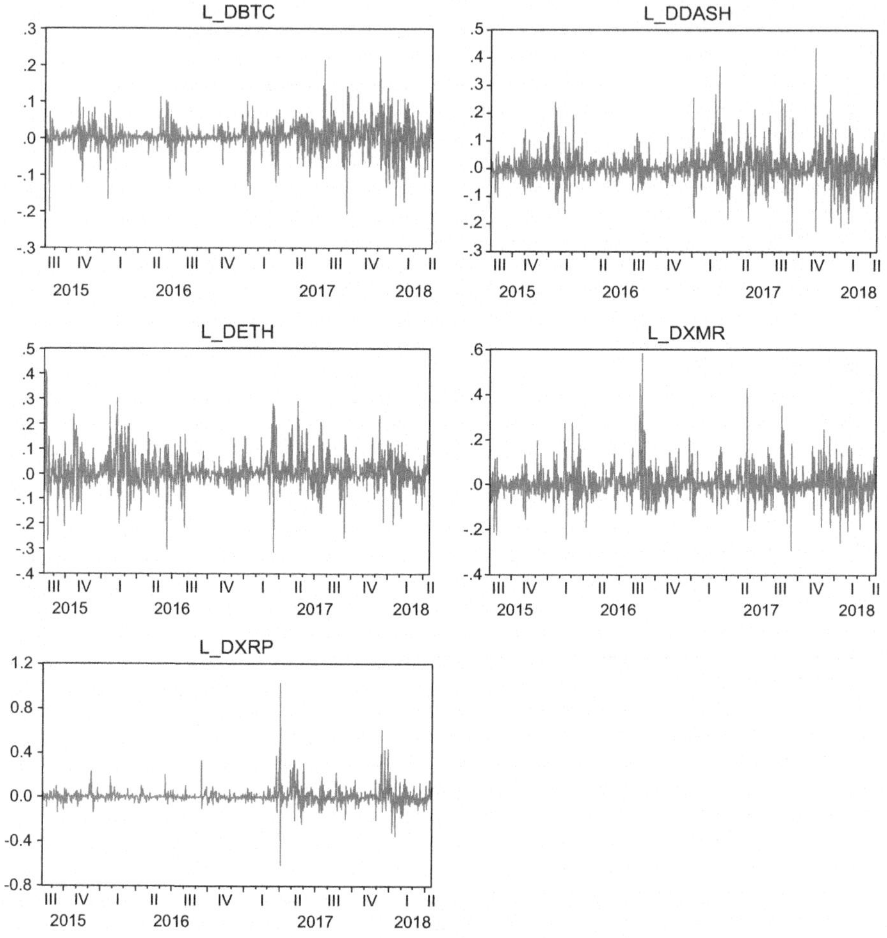

Fig. 2 Returns of prices from August 2015 to April 2018. *Note* The returns of L_DBTC, L_DETH, L_DDASH, L_DXMR and L_DXRP correspond to the logarithm first differencies to Bitcoin, Ethereum, Dash, Ripple and Monero respectively

4 Methodology

The variable of interest in this paper is the daily average returns of the cryptocurrency prices that are computed from the first differences of the natural logarithm of the five digital currencies [22]. From the features that were observed in the previous sections we concluded that a GARCH process is more appropriate for this research [8]. Our goal is to analyze the interdependence between the different digital currencies that we have included in this study. In order to achieve that, we will use a multivariate GARCH model [6] in the style of GARCH-CCC that was proposed by Bollerslev [5]. Therefore, because GARCH-CCC is a general formulation of the VECH model

Table 1 Summary statistics of the returns during from August 2015 to April 2018

	BTC	ETH	DASH	XRP	XMR
Mean	0.003493	0.006639	0.004859	0.004440	0.005764
Std. dev.	0.041377	0.072237	0.061343	0.079978	0.074643
Skeness	−0.263754	0.530203	0.929469	3.094100	1.083345
Kurtosis	7.896749	7.376236	9.109016	41.51131	10.20350
Jarqu-Bera	992.4917	829.6226	1668.407	62251.21	2315.267
Probability	0.000000	0.000000	0.000000	0.000000	0.000000

Note The prices of BTC, ETH, DASH, XRP and XMR correspond to Bitcoin, Ethereum, Dash, Ripple and Monero respectively. For normal distribution, skewness and kurtosis are 0 and 3 respectively. The critical values for Jarwue-Bera with 2° of freedom for significance of 10%, 5% and 1% are 4.61, 5.99 and 9.21 respectively

we will present firstly the VECH model. Bollerslev et al. [6] represented a common form of the VECH model.

$$VECH(H_t) = VECH(C) + \sum_{i=1}^{q} A_i VECH\left(\varepsilon_{t-1}\varepsilon'_{t-i}\right) + \sum_{i=1}^{q} B_i VECH(H_{t-1}) \quad (1)$$

$$\varepsilon_t|\psi_{t-1} \sim N(0, H_t),$$

Brooks [8] where H_t is an $N \times N$ conditional covariance matrix, ε_t is an $N \times 1$ innovation vector, ψ_{t-1} represents the information set at $t-1$ and $VECH(\cdot)$ refers to the column-stacking operator applied to the upper portion of the symmetric matrix. The unconditional variance of the VECH will be given by $C[I - A - B]^{-1}$ where I is an identity matrix of order $N(N+1)/2$.

The problem with VECH is that, if the number of parameters is large, then a set of parameter restrictions is needed to ensure that the conditional variance matrix H_t is positive definite. One method of reducing the number of parameters in the MGARCH framework is for the correlations between the disturbances, ε_t, to be fixed through the series of time [5] proposed the constant conditional correlation (CCC) model. In this model the conditional variances are identical to those of a univariate GARCH specification in the fixed correlation model. The CCC model is specified in two stages [8]. The first is the univariate GARCH specification model.

$$H_{ii,t} = C_i + A_i\varepsilon_{i,t-i} + B_i H_{ii,t-1}, \qquad i = 1, \ldots, N \quad (2)$$

in which the coefficient outputs are linear but they can be described with an N × 1 matrix for a better representation in our case.

$$H_{ij,t} = R_{ij} H_{ii,t}^{1/2} H_{jj,t}^{1/2}, \qquad i, j = 1, \ldots, N, i < j \tag{3}$$

The second stage of the constant is a conditional correlation model, that attempts to define the correlations of the off-diagonal elements of the $H_{ij,t}$, $(i \neq j)$; those are defined via the correlations that are denoted in the term R_{ij} on Eq. (3). A sufficient condition is for the $H_{ii,t}$ to have positive elements for all t with the C_i also defined positive with A_i and B_i elements for each i been non-negative. This guarantees that, together with the positive definiteness of R_{ij}, the conditional variance matrix $H_{ij,t}$ is positive definite almost certainly for all t. With the null hypothesis of no volatility interactions between our cryptocurrencies [33] we will try to find the linkages in the off-diagonal elements of the $H_{ij,t}$.

Since in our research we include five different digital currencies, therefore the matrix R_{ij} will be a 5 × 5 degree matrix. The unrestricted model that was described above is highly parameterized and it is very challenging to estimate. Therefore, we are using the constant conditional correlation (CCC) model for calculating our estimates [23]. In order to have a better understanding the elements can be defined as:

$$A_t = \begin{bmatrix} a_1 \\ a_2 \\ a_3 \\ \ldots \\ a_{983} \end{bmatrix}, B_t = \begin{bmatrix} b_1 \\ b_2 \\ b_3 \\ \ldots \\ b_{983} \end{bmatrix}, H_t = \begin{bmatrix} h_1 \\ h_2 \\ h_3 \\ \ldots \\ h_{983} \end{bmatrix}, \quad \forall j, j$$

$$H_{ij,t} = \begin{bmatrix} h_{1,1t} & h_{2,1t} & h_{3,1t} & h_{4,1t} & h_{5,1t} \\ h_{1,2t} & h_{2,2t} & h_{3,2t} & h_{4,2t} & h_{5,2t} \\ h_{1,3t} & h_{2,3t} & h_{3,3t} & h_{4,3t} & h_{5,3t} \\ h_{1,4t} & h_{2,4t} & h_{3,4t} & h_{4,4t} & h_{5,4t} \\ h_{1,5t} & h_{2,5t} & h_{3,5t} & h_{4,5t} & h_{5,5t} \end{bmatrix},$$

$$R_{ij} = \begin{bmatrix} 1 & \rho_{2,1} & \rho_{3,1} & \rho_{4,1} & \rho_{5,1} \\ \rho_{1,2} & 1 & \rho_{3,2} & \rho_{4,2} & \rho_{5,2} \\ \rho_{1,3} & \rho_{2,3} & 1 & \rho_{4,3} & \rho_{5,3} \\ \rho_{1,4} & \rho_{2,4} & \rho_{3,4} & 1 & \rho_{5,4} \\ \rho_{1,5} & \rho_{2,5} & \rho_{3,5} & \rho_{4,5} & 1 \end{bmatrix}, \quad \varepsilon_t = \begin{bmatrix} e_{1t} \\ e_{2t} \\ e_{3t} \\ e_{4t} \\ e_{5t} \end{bmatrix}$$

The CCC model that was described above can be estimated by maximizing the log-likelihood function. The log-likelihood function of the joint distribution is the sum of all individual log likelihood functions of the conditional distributions and the sum of the logs of the multivariable normal distribution.

$$l(\theta) = -\frac{TN}{2} \log 2\pi - \frac{1}{2} \sum_{i=1}^{T} (\log|H_t| + \varepsilon_t' H_t^{-1} \varepsilon_t) \tag{4}$$

where θ donates to all the unknown parameters to be estimated, N is the number of assets or in our case digital currencies and T is the number of observations and the rest notions are as above.

Berndt-Hall-Hall-Hausman (BHHH) algorithm follows the Newton-Raphon approach but replaces the negative of the Hessian by an approximation formed from the sum of the outer products of the gradient vectors for each observation's contribution to the objective function. As a numerical procedure in this study we use Marquardt's modification of the BHHH in which he adds a correction matrix to the outer product matrix.

5 Empirical Results

In this section, we are reporting the results of all estimations previously mentioned. We are looking for statistical significance across the results, so that we can find the effect of the individual past shocks and volatility of each digital currency (i) on its conditional variance [34]. In the last part we will show a generalized covariance demonstration in a line of VAR estimations from the CCC model as mentioned before.

Table 2 shows the results of the five-variable symmetric GARCH model converges after 53 iterations from Eqs. (2) and (3). To begin with, we are going to discuss the first step of the model, the univariate GARCH model. Coefficients $\left[a_{i,1}, a_{i,2}, \ldots, a_{i,5}\right]$ are results from the matrix A_t from Eq. (2) and as it is shown in Table 2 all of them are statistically significant at a 1% level. This shows that there is a strong ARCH effect, meaning there is a strong influence of the individual past shocks and volatility for every digital currency. Coefficients $\left[b_{i,1}, b_{i,2}, \ldots, b_{i,5}\right]$ are contained in the matrix B_t from Eq. (2) and again as it is shown in Table 2 all of them are statistically significant at a 1% level. Matrix B captures the GARCH effect and this again shows that there is a strong influence of the individual past shocks and volatility for every digital currency. We also observe that the GARCH effect is almost two times stronger than the ARCH effect, something that was expected from the statistical description of the series. Therefore, we argue that there is a strong GARCH(1, 1) process driving the conditional covariances of the five cryptocurrencies. Individual past shocks and volatility affect the conditional covariances of BTC, ETH, DASH, XRP and XMR.

Next, we want to analyze the time-varying conditional covariances that were estimated by the MGARCH-CCC [16]. However, before doing that, with our correct return data we also want to estimate the second stage of constant conditional correlation coefficients (R_{ij}) between of all five variables so we will have a first constant look, Table 3 presents the results with coefficients $\left[\rho_{i,1}, \rho_{i,2}, \ldots, \rho_{i,5}\right]$ that are elements from the matrix R_{ij} from Eq. (3). From the initial impression we see that all results are statistically significant at a 1% level. That indicates that we have the existence of a strong correlation between our variables. This, in its turn, is resulting in a high interdependent and

In Fig. 3, as mentioned before, we will present the time-varying conditional covari-

Table 2 Estimated coefficients for the five-variable MGARCH-CCC model

	BTC ($i = 1$)	ETH ($i = 2$)	DASH ($i = 3$)	XRP ($i = 4$)	XMR ($i = 5$)
$a_{i,1}$	0.132793*** (0.013086)	–	–	–	–
$a_{i,2}$	–	0.247303*** (0.023007)	–	–	–
$a_{i,3}$	–	–	0.238865*** (0.023310)	–	–
$a_{i,4}$	–	–	–	0.376311*** (0.025312)	–
$a_{i,5}$	–	–	–	–	0.181153*** (0.025926)
$b_{i,1}$	0.839231*** (0.013132)	–	–	–	–
$b_{i,2}$	–	0.716824*** (0.019491)	–	–	–
$b_{i,3}$	–	–	0.712085*** (0.024918)	–	–
$b_{i,4}$	–	–	–	0.885452*** (0.015918)	–
$b_{i,5}$	–	–	–	–	0.696781*** (0.036591)

Note The constants of the matrix C_i are not included. Values in the parenthesis are the standard errors, ***represent statistical significance at level of 1%

Table 3 Estimated coefficients for the five-variable MGARCH-CCC model

	BTC ($i = 1$)	ETH ($i = 2$)	DASH ($i = 3$)	XRP ($i = 4$)	XMR ($i = 5$)
$a_{i,1}$	1				
$a_{i,2}$	0.287511*** (0.025256)	1			
$a_{i,3}$	0.332126*** (0.025402)	0.361624*** (0.027900)	1		
$a_{i,4}$	0.278357*** (0.031053)	0.270235*** (0.027894)	0.237482*** (0.027715)	1	
$a_{i,5}$	0.416338*** (0.024881)	0.367759*** (0.027706)	0.420726*** (0.024681)	0.291949*** (0.025024)	1

Note The constants of the matrixes $H_{ii,t}^{1/2} H_{jj,t}^{1/2}$ are not included. In the matrix above $a_{i,j} = a_{j,i}$ so the upper-diagonal elements are not included. Values in the parenthesis are the standard errors, ***represent statistical significance at level of 1%. Volatile system of assets which will not fit into a low-risk diversified portfolio. We also observe that Ripple is the least Bitcoin-affected cryptocurrency with a coefficient of 0.27 and Monero is the most Bitcoin-affected with a coefficient of 0.41

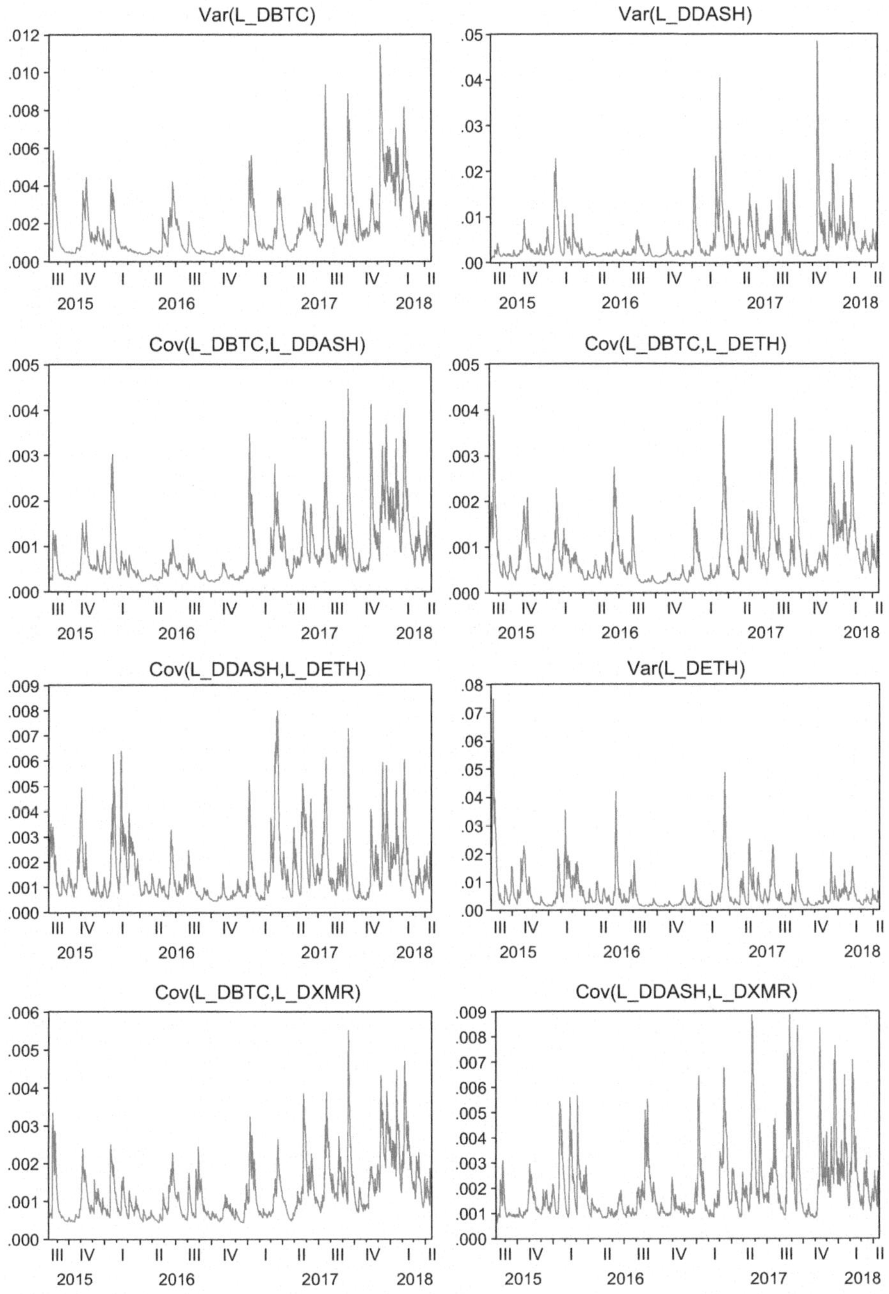

Fig. 3 Estimated conditional variance and conditional covariance. *Note* Where Var(i) is the variance of the variable i and were Cov(i, j) is the covariance between variables i and j, the returns of L_DETH, L_DDASH, L_DXMR and L_DXRP correspond to the logarithm first differencies to Bitcoin, Ethereum, Dash, Ripple and Monero respectively

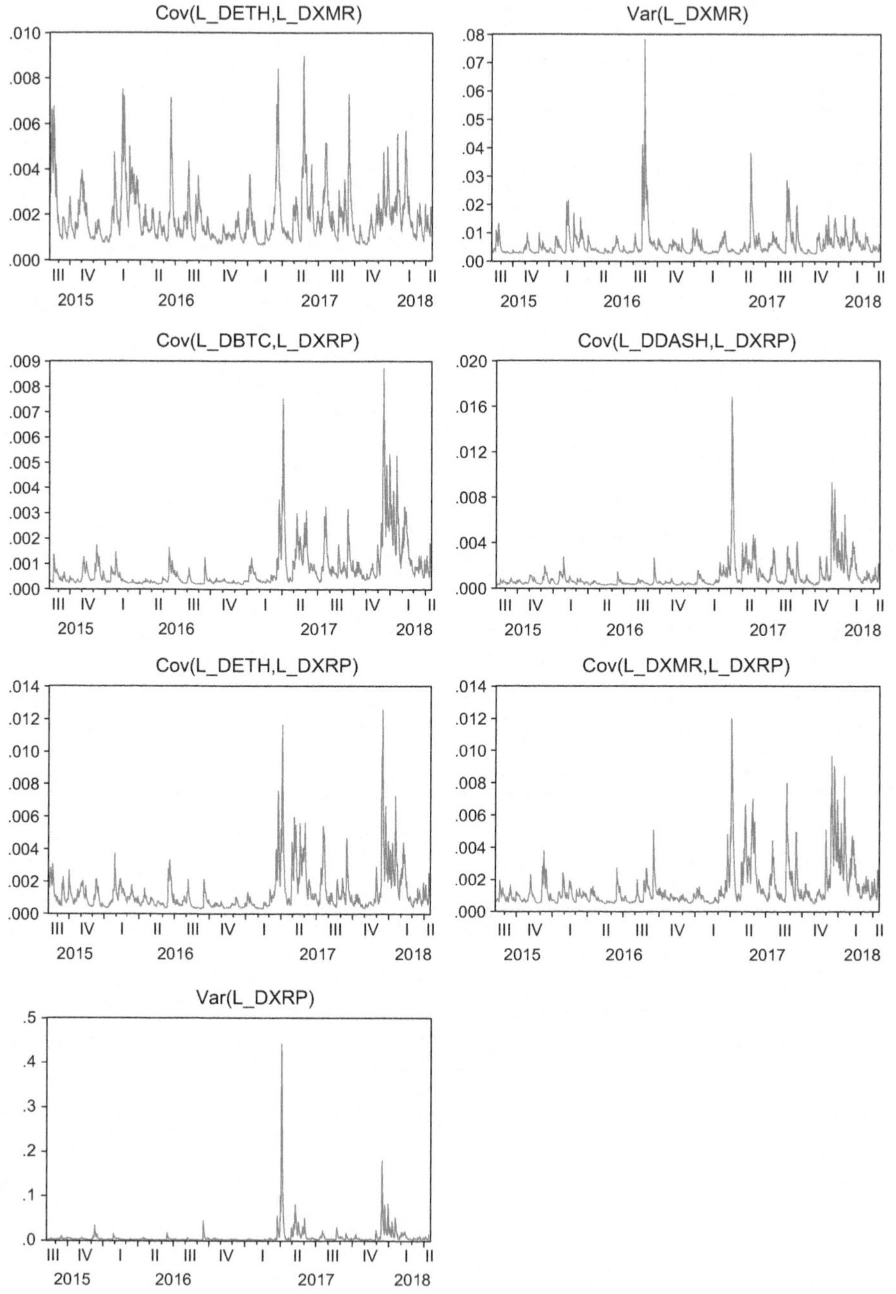

Fig. 3 (continued)

ances that were estimated by the MGARCH-CCC in a line of VAR. Our estimates confirm that we have positive values and conditional covariances that contain all individual variances and all covariance couples from the variables. We observe very similarly patterned results that confirm the previous observations that the system is very homogenous and it cannot effectively diversify risk [24]. The interesting part comes from the similarity of patterns that were observed when Bitcoin had its price apex between the end of 2017 and start of 2018, meaning that all different digital currencies covariances were severely affected from the price upraise of the Bitcoin [32].

6 Conclusion

In this paper we have attempted to establish quantifiable linkages between five different cryptocurrencies, those being Bitcoin, Ethereum, Ripple, Dash and Monero. Our literature review uncovered a significant lack of relevant research. This is due to the fact that cryptocurrencies, as well as the blockchain technology in general are a relatively new sector, and one that has not been clearly defined scientifically. Moreover, most existing studies use univariate approaches in their methodology.

In our attempt to bridge this gap and to fully address the complexity in a system of cryptocurrencies, we elected to use a multivariate approach instead. Consequently, we have shown that a very strong interlinkage and correlation does exist between a five-variable system of big capitalization cryptocurrencies, those being; Bitcoin, Ethereum, Ripple, Dash and Monero. We came up with the results using daily data from the period of 2015 to 2018, that were applied in the conditional asymmetric GARCH-CCC model. Our results were presented in Tables 2 and 3 and we could show their dynamic correlations between all the couples in Sect. 5.

Unfortunately, in our results we could not establish a clear causality from the impact that Bitcoin has due to its capital size compared to the whole market. But is very clear that the covariances of all the couples follow very similar patterns, meaning that the system of cryptocurrencies is very homogeneous and, as a result, diversification between those currencies will not significantly lower a portfolio's risk.

References

1. Baek, C., Elbeck, M.: Bitcoins as an investment or speculative vehicle? A first look. Appl. Econ. Lett. **22**(1), 30–34 (2015)
2. Balcilar, M., et al.: Can volume predict Bitcoin returns and volatility? A quantiles-based approach. Econ. Model. **64**, 74–81 (2017)
3. Bera, A.K., Kim, S.: Testing constancy of correlation and other specifications of the BGARCH model with an application to international equity returns. J. Empir. Financ. **9**(2), 171–195 (2002)

4. Blau, B.M.: Price dynamics and speculative trading in bitcoin. Res. Int. Bus. Financ. **41**, 493–499 (2017)
5. Bollerslev, T.: Modelling the coherence in short-run nominal exchange rates: a multivariate generalized ARCH model. Rev. Econ. Stat. **72**(3), 498–505 (1990)
6. Bollerslev, T., Engle, R.F., Wooldridge, J.M.: A capital asset pricing model with time-varying covariances. J. Polit. Econ. **96**(1), 116–131 (1988)
7. Bouoiyour, J., et al.: What drives Bitcoin price. Econ. Bull. **36**(2), 843–850 (2016)
8. Brooks, C.: Introductory Econometrics for Finance. Cambridge University Press (2019)
9. Chu, J., et al.: GARCH modelling of cryptocurrencies. J. Risk Financ. Manag. **10**(4), 17 (2017)
10. Conrad, C., Custovic, A., Ghysels, E.: Long-and short-term cryptocurrency volatility components: A GARCH-MIDAS analysis. J. Risk Financ. Manag. **11**(2), 23 (2018)
11. Cryptocurrency Market Capitalizations. CoinMarketCap, http://coinmarketcap.com/. Last Accessed 25 May 2018
12. Dash (Cryptocurrency): Wikipedia, Wikimedia Foundation, 14 Feb 2019. en.wikipedia.org/wiki/Dash_(cryptocurrency)
13. Dyhrberg, A.H.: Hedging capabilities of bitcoin. Is it the virtual gold? Financ. Res. Lett. **16**, 139–144 (2016)
14. Engle, R.F., Kroner, K.F.: Multivariate simultaneous generalized ARCH. Econom. Theory **11**(1), 122–150 (1995)
15. Ethereum: Wikipedia, Wikimedia Foundation, 14 Feb 2019. en.wikipedia.org/wiki/Ethereum
16. Ferenstein, E., Gasowski, M.: Modelling stock returns with AR-GARCH processes. SORT-Stat. Oper. Res. Trans. **28**(1), 55–68 (2004)
17. Gandal, N., Halaburda, H.: Competition in the Cryptocurrency Market. (2014)
18. Glaser, F., et al.: Bitcoin-asset or currency? Revealing users' hidden intentions. Revealing Users' Hidden Intentions, 15 April 2014. ECIS (2014)
19. Gronwald, M.: The Economics of Bitcoins–Market Characteristics and Price Jumps (2014)
20. Hafner, C., Franses, PHPHBF.: A generalized dynamic conditional correlation model for many asset returns. No. EI 2003-18 (2003)
21. Katsiampa, P.: Volatility estimation for Bitcoin: a comparison of GARCH models. Econ. Lett. **158**, 3–6 (2017)
22. Ku, Y.-H.H., Chen, H.-C., Chen, K.-H.: On the application of the dynamic conditional correlation model in estimating optimal time-varying hedge ratios. Appl. Econ. Lett. **14**(7), 503–509 (2007)
23. Ledoit, O., S-C, P., Wolf, M.: Flexible multivariate GARCH modeling with an application to international stock markets. Rev. Econ. Stat. **85**(3), 735–747 (2003)
24. Li, H., Majerowska, E.: Testing stock market linkages for Poland and Hungary: a multivariate GARCH approach. Res. Int. Bus. Financ. **22**(3), 247–266 (2008)
25. Monero (Cryptocurrency): Wikipedia, Wikimedia Foundation, 17 Feb 2019. en.wikipedia.org/wiki/Monero_(cryptocurrency)
26. Nakamoto, S.: Bitcoin: A Peer-to-Peer Electronic Cash System (2008)
27. Nakatani, T., Teräsvirta, T.: Testing for volatility interactions in the constant conditional correlation GARCH model. Econom. J. **12**(1), 147–163 (2009)
28. Panagiotidis, T., Stengos, T., Vravosinos, O.: On the determinants of bitcoin returns: a LASSO approach. Financ. Res. Lett. **27**, 235–240 (2018)
29. Panagiotidis, T., Stengos, T. Vravosinos, O.: The effects of markets, uncertainty and search intensity on bitcoin returns. Int. Rev. Financ. Anal. (2018)
30. Stavroyiannis, S., Babalos, V.: Dynamic Properties of the Bitcoin and the US Market (2017)
31. Swan, M.: Blockchain thinking: the brain as a dac (decentralized autonomous organization). Texas Bitcoin Conference, Chicago (2015)
32. Tse, Y.K.: A test for constant correlations in a multivariate GARCH model. J. Econom. **98**(1), 107–127 (2000)
33. Tsui, A.K., Qiao, Y.: Constant conditional correlation in a bivariate GARCH model: evidence from the stock markets of China. Math. Comput. Simul. **48**(4-6), 503–509 (1999)

34. Varga-Haszonits, I., Kondor, I.: Noise sensitivity of portfolio selection in constant conditional correlation GARCH models. Phys. A Stat. Mech. Appl. **385**(1), 307–318 (2007)
35. Vasilios, K., Athanasios, Z., Papangelou, S.: Small forensic smart-law-scripts the first step for intelligent justice punishment in law enforcement, economic crime and alternative sentences. Bus. Econ. Res. **8**(2), 154–167 (2018)

Compact Storage of Superblocks for NIPoPoW Applications

Kostis Karantias, Aggelos Kiayias and Dionysis Zindros

Abstract Blocks in proof-of-work (PoW) blockchains satisfy the PoW equation $H(B) \leq T$. If additionally a block satisfies $H(B) \leq T2^{-\mu}$, it is called a *μ-superblock*. Superblocks play an important role in the construction of compact blockchain proofs which allows the compression of PoW blockchains into so-called *Non-Interactive Proofs of Proof-of-Work* (NIPoPoWs). These certificates are essential for the construction of *superlight* clients, which are blockchain wallets that can synchronize exponentially faster than traditional SPV clients. In this work, we measure the distribution of superblocks in the Bitcoin blockchain. We find that the superblock distribution within the blockchain follows expectation, hence we empirically verify that the distribution of superblocks within the Bitcoin blockchain has not been adversarially biased. NIPoPoWs require that each block in a blockchain points to a sample of previous blocks in the blockchain. These pointers form a data structure called the *interlink*. We give efficient ways to store the interlink data structure. Repeated superblock references within an interlink can be omitted with no harm to security. Hence, it is more efficient to store a *set* of superblocks rather than a *list*. We show that, in honest executions, this simple observation reduces the number of superblock references by approximately a half in expectation. We then verify our theoretical result by measuring the improvement over existing blockchains in terms of the interlink sizes (which we improve by 79%) and the sizes of succinct NIPoPoWs (which we improve by 25%). As such, we show that deduplication allows superlight clients to synchronize 25% faster.

K. Karantias
University of Ioannina, Ioannina, Greece

A. Kiayias
University of Edinburgh, Edinburgh, Scotland

D. Zindros (✉)
University of Athens, Athens, Greece
e-mail: dionyziz@gmail.com

A. Kiayias · D. Zindros
IOHK, Edinburgh, Scotland

P. Pardalos et al. (eds.), *Mathematical Research for Blockchain Economy*,
Springer Proceedings in Business and Economics,
https://doi.org/10.1007/978-3-030-37110-4_6

1 Introdution

Bitcoin [1] and other blockchains which use the same backbone consensus mechanism [2] use Simple Payment Verification (SPV) to shorten the synchronization time for lightweight clients, where the clients need to download block headers instead of whole blocks. Recently, a line of work has introduced *superlight* clients, which do not require all blockchain headers to be downloaded, but, rather, only a sample of them. This sample consists of blocks which happen to achieve a higher difficulty than the required one, and are thus termed *superblocks*.

By sampling the superblocks of a chain, short proofs about a blockchain can be created, which allow a client to synchronize with the longest blockchain without downloading all blocks. These so-called *proofs of proof-of-work* contain only a small number of cleverly chosen superblocks which compact the proof-of-work of the blockchain into a succinct string, while maintaining the same security level as SPV clients. However, while the protocol has even been deployed in practice, the distribution of superblocks within a blockchain has not been previously measured. In this paper, we provide measurements of this distribution for the Bitcoin blockchain.

In order for superblock sampling to work, it is necessary that each block contains, in addition to the standard pointer to its previous block, a few select pointers to some preceding superblocks. These pointers are organized in a special data structure, the *interlink*. For relevant applications such as superlite clients and cross-chain transfers, it is critical that the interlink structure is compact. We measure the size of the interlink structure and provide a simple novel optimization which can bring down its size to less than a half. We then study the impact of this improvement on the size of proofs of proof-of-work.

Related work. Superblocks were first observed to exist in [3]. The interlink data structure was put forth in [4], where it was also observed that it can be organized into a Merkle tree. Interlinks containing all the blocks of the blockchain have been proposed in [5]. Superblock interlinks have been included from genesis in cryptocurrencies such as ERGO [6] and nimiq [7]. Complete blockchain interlinks have been proposed for Ethereum [8]. Nimiq and ERGO have independently applied interlink deduplication in practice to save space [6, 9]. In [10], the consumption of the interlink data to construct Non-Interactive Proofs of Proof-of-Work was presented and concrete numbers were given about the sizes of such proofs. They also presented a way to construct such a structure without a soft or hard fork, but a backwards compatible *velvet fork*, which was later explored in [11]. Bitcoin Cash has been velvet forked in this manner [12]. Beyond superlight clients, another application of NIPoPoWs are cross-chain transfers [13] between proof-of-work blockchains. Comparable constructions have also appeared for proof-of-stake blockchains [14].

Our contributions. The contributions of this paper are summarized as follows:

1. We measure superblock distributions in Bitcoin. We observe that the distribution of superblocks follows expectation, indicating there are no ongoing or historical attempts to bias the distribution of superblocks (so-called *badness* attacks [10]).

We are the first to collect any empirical measurements of superblocks on real blockchains.
2. We describe the simple but important optimization in regards to the way blocks are compactly stored in an interlink tree by observing that duplicate pointers can be removed without harming security. As such, we construct interlink *block sets* instead of interlink *block lists*.[1]
3. We prove that our optimization reduces the number of pointers in each interlink by a half in expectation.
4. We evaluate our improvement on the Bitcoin blockchain and collect empirical data regarding the performance of our improvement, including concrete sizes of NIPoPoWs built. We experimentally demonstrate that our optimization reduces interlink vector sizes by 79% on average and the already very succinct NIPoPoW certificates by 25% on average.

2 Superblocks and Proofs-of-Proofs

Blocks generated in proof-of-work [15] systems must satisfy the proof-of-work equation $H(B) \leq T$ where T denotes the mining target [16] and B denotes the block contents, which is a triplet including a representation of the application data and metadata, a nonce, and a reference to the previous block by its hash. The function H is a hash function, modelled as a random oracle [17], which outputs κ bits, where κ is the security parameter of the protocol and $T < 2^{\kappa}$. It sometimes happens that some blocks satisfy a stronger version of the equation [4], namely that $H(B) \leq T2^{-\mu}$ for some $\mu \in \mathbb{N}$. Such blocks are called μ-superblocks [10]. It follows directly from the Random Oracle model that $\Pr[B \text{ is a } \mu\text{-superblock} | B \text{ is a valid block}] = 2^{-\mu}$. Note that if a block is a μ-superblock for some $\mu > 0$, then it is also a $(\mu - 1)$-superblock. We denote the maximum μ of a block B its $level(B) = \lfloor \lg(T) - \lg(H(B)) \rfloor$.

The count of superblocks in a chain decreases exponentially as μ increases. If a blockchain $\mathcal{C}$ generated in an honest execution has $|\mathcal{C}|$ blocks, it only has $2^{-\mu}|\mathcal{C}|$ superblocks of level μ in expectation. Hence, the total number of levels is $\lg(|\mathcal{C}|)$ in expectation. It has been theoretically posed that the distribution of superblocks can be adversarially biased in so-called "badness" attacks [10] in which an adversary reduces the density of superblocks of a particular level within a blockchain. However, the actual distribution of superblocks in currently deployed blockchains has not been measured. Therefore, it was previously unknown whether such attacks are taking place in the wild. In this paper, we make empirical measurements of superblock distributions and observe that they follow the expectation. Hence, we conclude that widespread badness attacks have not occurred in practice, confirming previous suspicions that such attacks are costly to mount.

[1] The deduplication optimization has already been discovered and deployed independently by the Nimiq and the ERGO blockchains [6, 9], but with no further analysis.

For any block B, it is useful to be able to refer to its most recent preceding μ-superblock for any $\mu \in \mathbb{N}$. In addition, it is useful to include this reference within the contents of the block to which proof-of-work is being applied so that the miner proves that she had knowledge of the preceding superblock when B was generated. For this purpose, it has been recommended [10] that for each block B, instead of including only a pointer to the previous block, $\lg(|\mathcal{C}|)$ pointers will be included, one for each level μ pointing to the most recent μ-superblock preceding B. Hence, under this modification, every block contains a pointer to its most recent 0-superblock ancestor, its most recent 1-superblock ancestor, and so on, of which there are $\lg(|\mathcal{C}|)$. These pointers change the blockchain into a block skiplist [18, 19].

These $\lg(|\mathcal{C}|)$ pointers per block are called the *interlink*. One way to include them is to replace the previd pointer, which in typical blockchains points to the previous block hash, with the interlink list of block hashes to be included *verbatim* in the block header. Alternatively, the interlink list of hashes can be organized into a compact data structure such as a Merkle tree [20] containing one leaf per superblock level μ. The number of leafs in this Merkle tree is $\lg(|\mathcal{C}|)$ and its height is $\lg\lg(|\mathcal{C}|)$. Hence, proofs-of-inclusion in this Merkle tree are of size $\Theta(\lg\lg(|\mathcal{C}|))$. The root of this Merkle tree can be included in the block header, replacing previd. This is done in blockchains adopting interlinking from genesis or through a hard fork [6, 7].

More commonly, to avoid modifying the block header format, the interlink Merkle tree root can be included in the block's application data. In this case, the root of the Merkle tree appears as auxiliary data within a particular transaction which is included in the block. If the miners of the blockchain are aware of the interlink, then it can be required that they included it in their coinbase transaction. The veracity of the interlink data does not need to be verified when it is included in a block, as invalid or malicious data in the interlink does not harm security. Hence, it is possible to include the interlink data in a user transaction. In this case, the transaction which includes the root of the interlink is called a *velvet transaction* and its inclusion is termed a *user-activated velvet fork* [12]. In practice, this transaction is implemented using an OP_RETURN [21] committing to the Merkle tree root containing the interlink list in its leafs. User-activated velvet forks allow the adoption of a new rule without requiring miners to upgrade their software or be aware of the change, and are hence backwards-compatible.

It is useful to be able to prove that a block B contains a pointer to a particular ancestor B' in its interlink. This statement is proven by a full node who holds all blockchain data, the *prover* P, to a superlight *verifier* V who holds only the header of block B. This proof is straightforward. The header of block B contains the Merkle tree root of the transactions tree mtr_1 and is hence known to V. First, a Merkle tree proof-of-inclusion π_1 proves that mtr_1 contains the velvet transaction tx. The velvet transaction tx commits to auxiliary data which includes the interlink Merkle tree root mtr_2. Secondly, another Merkle tree proof-of-inclusion π_2 proves that mtr_2 contains the hash of B'. While we cannot improve the size of π_1, in this paper we describe the improvement in the size of π_2.

Superblock pointers can be used to traverse the blockchain from the tip back to Genesis in a manner which skips some unnecessary intermediary blocks and includes others. The idea is to convince a superlight verifier V, which only has access to the Genesis block, that a particular blockchain is the longest one without presenting all block headers. Blocks of interest that are part of the longest chain can then be revealed to V in order to convince them that a particular transaction has been confirmed. In order to do that, P finds a succinct sample of blocks and places it in chronological order. That sample is chosen such that each next block within the sample contains a pointer to its immediate ancestor within the sample by a commitment in the interlink vector. The prover P sends each block of the sample to V, along with a proof-of-inclusion for the respective pointer. The verifier V can check if the correct pointer has been included. By cleverly choosing which blocks to collect, a full node can prove to a superlight node that the currently adopted longest blockchain is the claimed one without presenting the whole blockchain. Hence, instead of transmitting data linear in the chain size $\Theta(|\mathcal{C}|)$ as SPV clients do, it is sufficient to transmit succinct certificates which are only of size $\Theta(poly \log(|\mathcal{C}|))$. Such certificates are called Non-Interactive Proofs of Proof-of-Work [10]. In this paper, we are not concerned about the mechanism by which the NIPoPoWs protocol samples blocks, but only the number of blocks in these samples and the sizes of their proofs-of-inclusion, which we optimize here.

The NIPoPoWs protocol is parameterized by a security parameter m. The number of blocks $|\pi|$ in a given Non-Interactive Proof of Proof-of-Work sample is as follows. For each superblock level $\mu \in \mathbb{N}$, consider the blocks in the honestly adopted blockchain $\mathcal{C}$. Among these, some are of level μ, so denote the count of μ-level superblocks in $\mathcal{C}$ as $|\mathcal{C}{\uparrow}^{\mu}|$. If $|\mathcal{C}{\uparrow}^{\mu}| \geq m$, then we call μ an *included level*. Consider the maximum included level $\max \mu$. It has been proven [10] that $\max \mu = \lg(|\mathcal{C}|) - \lg(m)$. The proof π contains $1.5m$ blocks for the maximum included level and m additional blocks for each lower level in expectation. Hence the number of blocks in a NIPoPoW is $|\pi| = 1.5m + m \max \mu = 1.5m + m(\lg(|\mathcal{C}|) - \lg(m))$. For each of these blocks, the proof contains the block hash and the respective proofs-of-inclusion for the pointer to the preceding ancestor. In this paper, we optimize the size of these proofs-of-inclusion, which gives a direct improvement to the size of such proofs π. We note that, in our proposed construction, we do not decrease the number of required blocks in π, only the bytes that need to be transmitted for it on the network.

The size of these proofs is critical. As the majority of the time needed for mobile wallets to perform the initial synchronization with the network is spent on downloading block headers from the network, bringing down the proof size directly improves the performance of superlight clients. In the context of cross-chain transfers [13], these proofs are posted and persistently stored in smart contracts [22, 23] within blockchains which function as SPV verifiers for other blockchains [24]. Improving their size has direct financial impact on the protocol, as a larger size incurs a larger gas cost for storage purposes in case such proofs are stored within Ethereum.

3 Superblock Distributions in Deployed Cryptocurrencies

We measured the superblock distribution in the mainnet Bitcoin blockchain. Our results are illustrated in Fig. 1. As expected, half the blockchain blocks are 1-superblocks, 1/4 of blocks are 2-superblocks and generally approximately $2^{-\mu}$ of the blockchain blocks are μ-superblocks. The horizontal axis denotes the block height, while the vertical axis denotes the superblock density with respect to the variable difficulty target of each block, in logarithmic scale.

We performed these measurements as follows. We downloaded the whole bitcoin blockchain from the Genesis block up to the current tip of the blockchain (at the time of writing 563,451). We then plotted the density for each level $\mu = 0, \ldots, 6$. For the particular level, we traversed the blockchain using a sliding window of $1000(2^{\mu}) + 1$ blocks. Within that sliding window, we measured how many blocks of level μ exist, and plotted the ratio of the count of these superblocks within the window to the window size. The plot for level 0 is flat, as all blocks are 0-superblocks. The high-frequency erratic behavior is due to the probabilistic nature of block generation. We conjecture that lower frequency patterns, especially those aligned between multiple levels, are due to difficulty adjustment which incorrectly predicted the underlying computational power for a given epoch (e.g., due to rapidly changing costs in mining hardware or cryptocurrency prices).

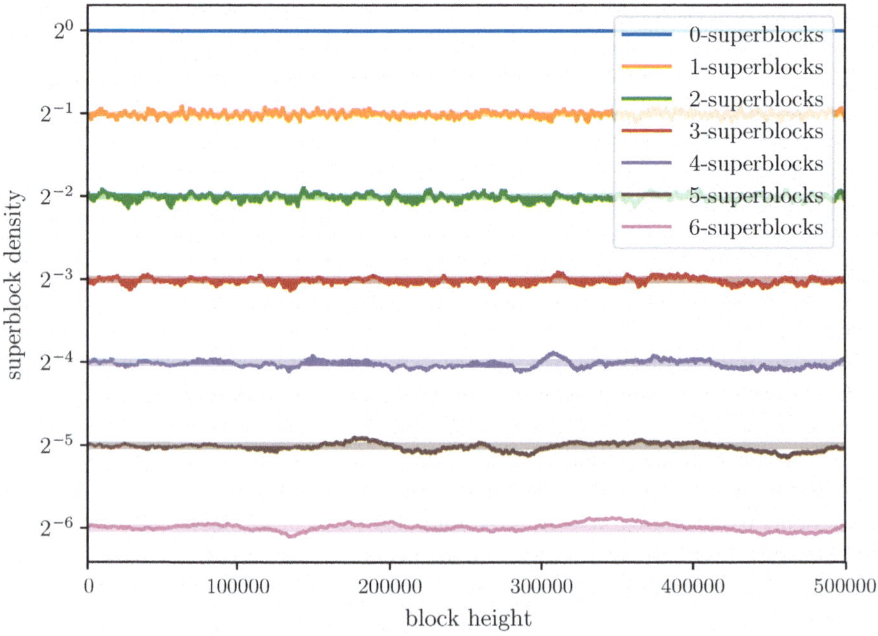

Fig. 1 Distribution of block levels in Bitcoin. Superblocks are with respect to variable difficulty targets. Shaded area indicates $\pm 1\sigma$

4 Interlinks as Sets of Superblocks

In superblock-enabled blockchains, the interlink vector stored in each block B contains one pointer per superblock level μ, namely a pointer to the most recent superblock preceding B of the respective level μ. This construction, known as an *interlink list*, is realized by inductively updating the interlink of the previous block, as shown in Algorithm 1. The algorithm works as follows. Trivially, genesis has an empty interlink vector, which forms our inductive basis. Given a newly mined block B' which already has an interlink vector (the inductive hypothesis), we wish to construct the interlink vector to be included in the next block, which will point to B' itself as well as some of the blocks that B' points to. This is done by inspecting the existing interlink, $B'.\mathsf{interlink}$, and constructing a new interlink $\mathsf{interlink}$ by replacing all the entries in $B'.\mathsf{interlink}$ that are of level less than or equal to that of B' with B' itself.

Algorithm 1 updateInterlink

```
function updateInterlink(B')
    interlink ← B'.interlink
    for μ = 0 to level(B') do
        interlink[μ] ← H(B')
    end for
    return interlink
end function
```

Algorithm 2 Our proposed algorithm, updateInterlinkSet

```
function updateInterlinkSet(B')
    interlinkSet ← {H(B')}
    for H(B) ∈ B'.interlink do
        if level(B) > level(B') then
            interlinkSet ← interlinkSet ∪ {H(B)}
        end if
    end for
    return interlinkSet
end function
```

Here, we make the simple observation that the interlink structure constructed in this manner often contains *duplicate pointers*. In fact, as we will show, most of the interlink pointers are duplicate. Space can be saved by constructing an *interlink set* instead. This construction is shown in Algorithm 2. The algorithm returns the exact same data structure as Algorithm 1, but with duplicates removed. The algorithm operates as follows. Given an existing interlink set, $B'.\mathsf{interlinkSet}$, it produces a new set $\mathsf{interlinkSet}$ which contains B' and all the same blocks as $B'.\mathsf{interlinkSet}$ with the exception of those that are of equal or inferior superblock level to B'.

Naturally, when this interlink set is to be committed to a Merkle tree, it must be ordered in a canonical matter (for example, by increasing block level) so that its root can be deterministically reproduced and detected. This canonical ordering may now not be trivial as was in the case for interlink lists and must be specified by the implementation.

We remark that it does not matter for security purposes whether duplicates are removed. The reason is that the prover has access to the whole list of blocks references within the interlink Merkle tree, and hence can choose the one it needs. On the other hand, the verifier only needs to ensure that the claimed superblock level is attained, but this can be done directly by inspecting the hash sent to it by the prover.

We now analyze the savings attained by the above method. We first analyze the savings in a thought experiment of an ideal, deterministic execution of the blockchain protocol. While this setting is not realistic, it provides good intuition about the interlink structure. Subsequently, we analyze the real probabilistic blockchain protocol. Consider a blockchain of n blocks.

Definition 1 Define the *interlink mask* of a block B to be the bitstring containing one bit per superblock level μ. At the position μ, the bitstring contains a 1 if the most recent μ-superblock ancestor of B differs from the most recent $(\mu + 1)$-superblock ancestor of B, or if no $(\mu + 1)$-superblock ancestor exists. Otherwise, it contains a 0.

This mask contains a 0 at the position of duplicates which can be eliminated. To measure the efficiency of our optimization scheme, we wish to count how many 0s are contained in the interlink mask of a given block. Consider, for example, the block highlighted with a dashed border in Fig. 2. Its interlink vector will have an interlink mask of 0101. The first 0 is due to the previous block which happened to be a 1-superblock. The latter 0 is due to the most recent 3-superblock overshadowing the preceding 2-superblock.

In our deterministic thought experiment, consider a blockchain which grows as illustrated in Fig. 2. In this blockchain, every block is a 0-superblock, every other block is a 1-superblock, every fourth block is a 2-superblock and generally every 2^μ-th block is a μ-superblock. In this thought experiment, the interlink mask behaves like a binary number which is increased by 1 after every block is generated. As such, it will be a μ-digit binary number. As the process passes through all μ-digit binary numbers, the number of 0s and 1s is on average equal. Hence, the savings obtained in the deterministic case are exactly 50%.

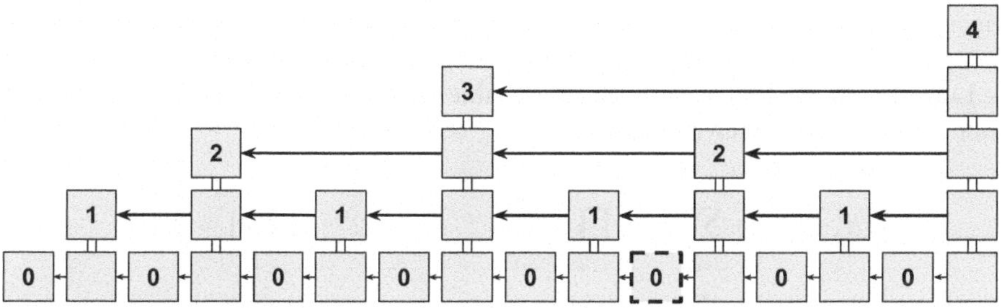

Fig. 2 A thought experiment of a deterministically generated blockchain

Algorithm 3 The Turing Machine modeling interlink generation.

function run(tape)
  $\mu \leftarrow 0$
  while true **do**
    $b \xleftarrow{\$} \{0, 1\}$
    if b = 1 **then**
      tape[μ] $\leftarrow 1$
      return
    end if
    tape[μ] $\leftarrow 0$
    $\mu \leftarrow \mu + 1$
  end while
  return tape
end function

Consider now the probabilistic setting of a real blockchain in an honest execution, where each block generated has an independent probability of belonging to a given level. The process of block generation can be modelled precisely as follows. Consider the Turing Machine illustrated in Algorithm 3. We begin with a one-sided infinite tape filled with the special symbol $\sqcup$. We then run the machine illustrated in Algorithm 3 repeatedly over the same tape n times. Once we have completed the n runs, the tape contains a binary string, which follows the same distribution as the interlink mask of the nth block of a blockchain. Each run of Algorithm 3 corresponds to a generation of a block. The algorithm begins at the position $\mu = 0$ of the tape. It flips a fair coin b. If the coin turns out to be 1, the machine writes 1 to the current position of the tape and exits. This is the event that the block generated has level exactly 0. Otherwise, if the coin b is a 0, then the block generated has level above 0, and so the first position of the interlink mask will be overwritten by a 0. The machine then advances and continues to flip coins and overwriting the tape with 0s until a 1 coinflip is attained, at which point it writes a 1 and halts. The probability of the machine halting at position μ or later is $2^{-\mu}$, modeling the probability of a block being a μ-superblock. The machine overwrites with 0 the positions in the tape which are of inferior level compared to the block level it will generate at the given run. This is the same process by which superblocks of higher level overshadow preceding superblocks of lower levels by occupying their space with duplicate pointers that can be eliminated.

Let $B^n_\mu \in \{0, 1\}$ denote the random variable containing the value of the μth digit after n such runs. We have that:

$$\Pr[B^n_\mu = 1] = \sum_{i=1}^{n} 2^{-\mu} \prod_{j=i+1}^{n} \sum_{\mu'=1}^{\mu-1} 2^{-\mu'} = \sum_{i=1}^{n} 2^{-\mu} \prod_{j=i+1}^{n} (1 - 2^{1-\mu})$$

$$= \sum_{i=1}^{n} 2^{-\mu}(1 - 2^{1-\mu})^{n-(i+1)+1} = \frac{1}{2}(1 - (1 - 2^{1-\mu})^n)$$

Let B^n denote the number of 1s in the interlink mask after n runs. Its expectation is then

$$\mathbb{E}[B^n] = \mathbb{E}\left[\sum_{\mu=1}^{\infty} B^n_\mu\right] = \sum_{\mu=1}^{\infty} \frac{1}{2}(1 - (1 - 2^{1-\mu})^n) = \frac{1}{2}\sum_{i=1}^{n}(-1)^{i+1}\binom{n}{i}\frac{2^i}{2^i - 1}$$

This series converges to a $\Theta(\lg(n))$ function which is close to $0.57 \cdot \lg(n)$ as illustrated in Fig. 3, indicating savings of approximately 43%.

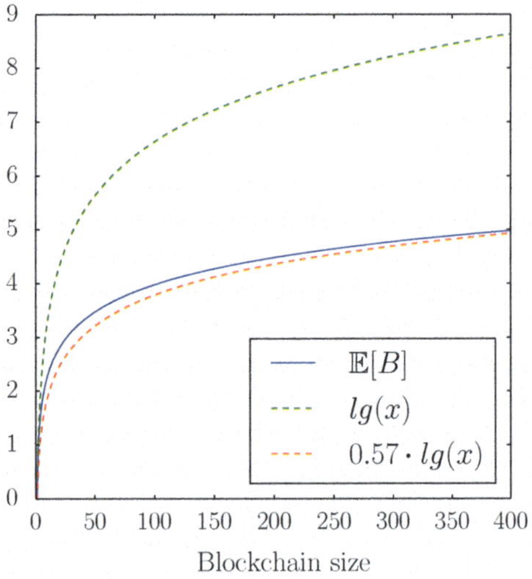

Fig. 3 The expected number of unique interlink pointers in a block

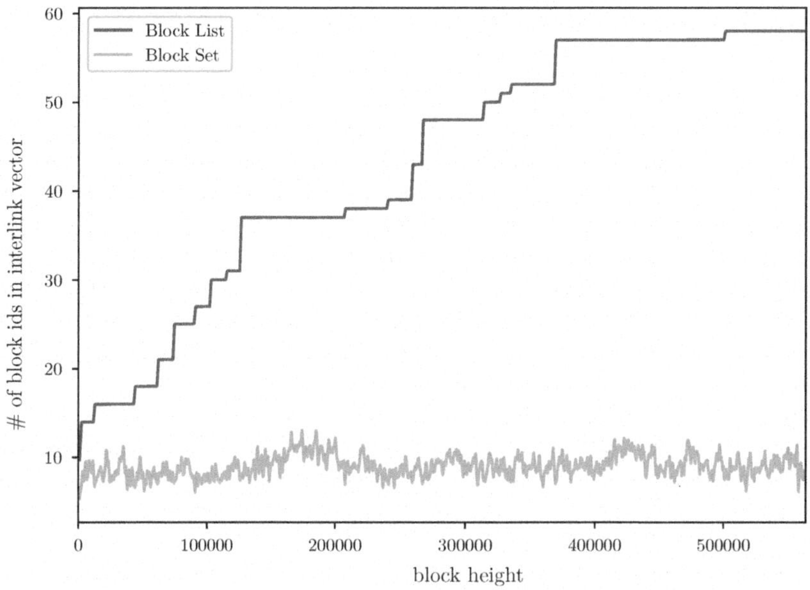

(a) Bitcoin

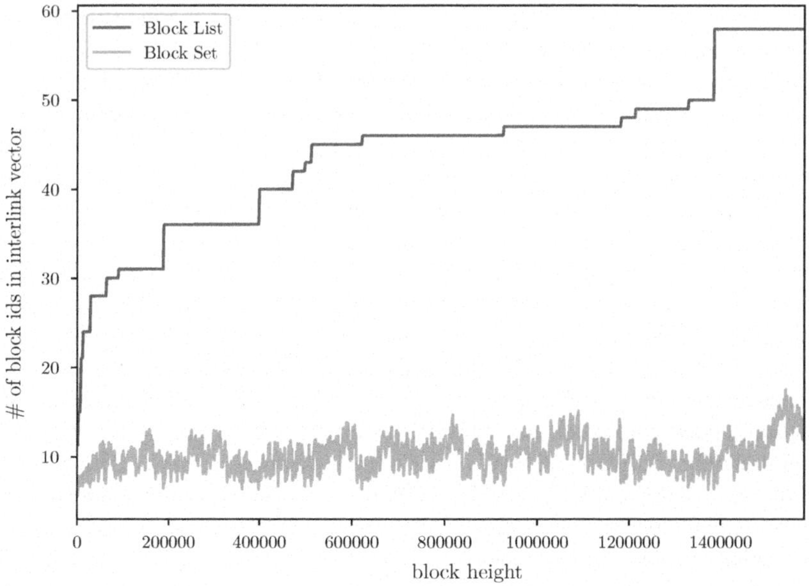

(b) Litecoin

Fig. 4 A comparison of interlink vector sizes for interlink block lists and interlink block sets in two popular blockchains (lower is better)

5 Empirical Analysis of Improvement

In order to empirically assess the space efficiency of our improvement, we measured the size of the interlink data structure in the case of interlink *lists*, the previously proposed format, and in the case of interlink *sets*, our newly proposed format. We performed our measurements on the mainnet for both Bitcoin and Litecoin. Our results are illustrated in Fig. 4 and are similar for both of these coins. The figures assume that both coins have been velvet forked from their genesis blocks to include the particular interlink vector format. This is indicative of the future performance of velvet forking each blockchain to add the respective interlink vector format.

The new data structure format yields savings of approximately 79% on average. Based on the theoretical analysis of Sect. 4, we expect to see approximately an improvement of 50% in this structure. The extra 29% is due to the historical explosion of difficulty in the mining power in both cryptocurrencies. The increased difficulty causes a lower variable difficulty target, meaning that the lowest portions of the superblock levels remain unoccupied, but are still accounted for in the interlink vector list approach.

Based on the sizes attained in the interlink vector of the Bitcoin blockchain, we organized the interlink vector into Merkle trees for both the list and the set structure and created proofs-of-inclusion of which we measured the size. The sizes of the inclusion proofs for the two constructions are illustrated in Fig. 5, while the percentile savings are illustrated in Fig. 6.

We summarize the savings of our construction in Table 1. The table was constructed by inspecting the Bitcoin blockchain at the time of writing. The interlink size column shows the average interlink vector size, in the number of block hashes and in concrete bytes assuming the `SHA256` function is used (as in Bitcoin). The proof-of-inclusion size column shows the average size of a Merkle proof-of-inclusion, in the number of hashes and in bytes, when the interlink vector is compacted into a Merkle tree using `SHA256`. Finally, the NIPoPoW size column shows the size of a NIPoPoW in kilobytes (excluding the last k blocks of the chain which must be sent *verbatim* irrespectively of which synchronization protocol is used). The NIPoPoW sizes are calculated assuming Bitcoin had included the respective interlink Merkle tree root in their headers since genesis. We measured the size of suffix-proof NIPoPoWs based on the recommended parameter $m = 15$ [10] assuming a chain size of $|C| = 563{,}451$. The number of blocks in a NIPoPoW is $(m(\lg|C| - \lg m) + 1.5m) = 250$ in expectation. For the final size calculation of the NIPoPoW, we included all the data required: The proofs-of-inclusion (based on the size given on the previous column) and the block headers needed (80 bytes per block). Our results indicate 79% savings in the interlink sizes and 25% savings in the NIPoPoW sizes.

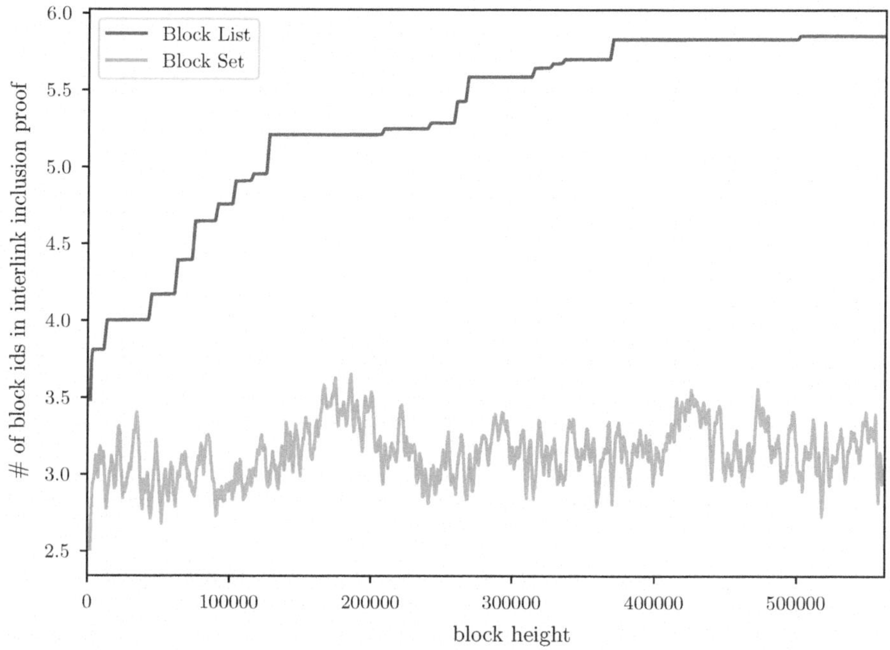

Fig. 5 A comparison of a proof-of-inclusion size in the case of interlink block lists and interlink block sets in Bitcoin (lower is better)

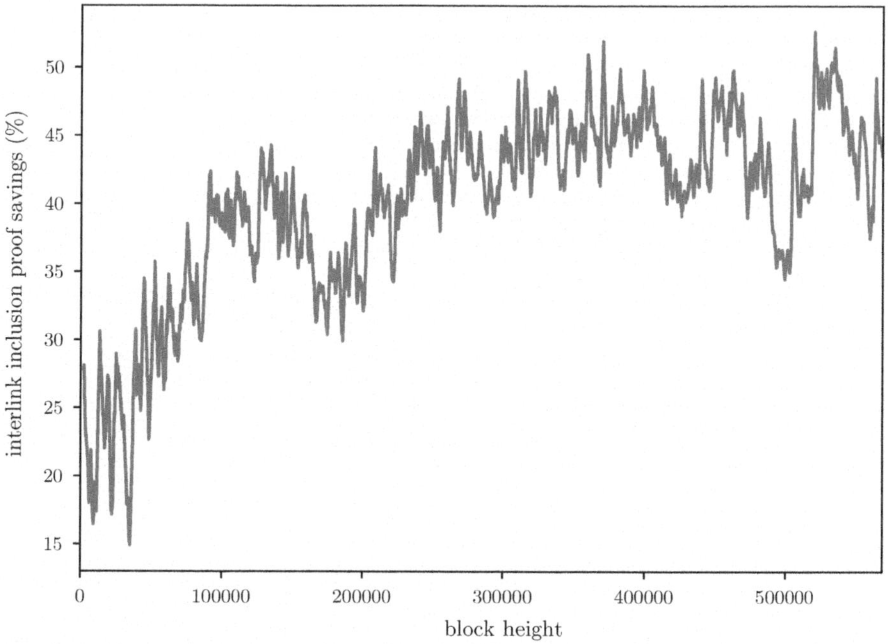

Fig. 6 Percentile savings of the block set construction compared to the block list construction

Table 1 A comparison of the two interlink constructions in terms of size

	Interlink size		Proof-of-inclusion size		NIPoPoW size
	Blockhashes	Bytes	Hashes	Bytes	KB
Interlink lists	43.12	1380	5.7	183	65.7
Interlink sets	9.04	289	3.6	116	49

References

1. Nakamoto, S.: Bitcoin: a peer-to-peer electronic cash system. https://bitcoin.org/bitcoin.pdf (2008)
2. Garay, J.A., Kiayias, A., Leonardos, N.: The bitcoin backbone protocol: analysis and applications. In: Oswald, E., Fischlin, M. (eds.) EUROCRYPT 2015. Part II, volume 9057 of LNCS, pp. 281–310. Springer, Heidelberg (2015)
3. Miller, A.: The high-value-hash highway, bitcoin forum post (2012)
4. Kiayias, A., Lamprou, N., Stouka, A.-P.: Proofs of proofs of work with sublinear complexity. In: International Conference on Financial Cryptography and Data Security, Springer, pp. 61–78 (2016)
5. Bünz, B., Kiffer, L., Luu, L., Zamani, M.: FlyClient: super-light clients for cryptocurrencies. Cryptology ePrint Archive, Report 2019/226. https://eprint.iacr.org/2019/226 (2019)
6. Chepurnoy, A., Meshkov, D., Oskin, I., Aksarin, M., Andreev, A., Slesarenko, A., Zadorozhnyi, D., Manzanares, G.: Ergo. https://ergoplatform.org (2018)
7. Chin, E., von Styp-Rekowsky, P., Linus, R.: Nimiq. https://nimiq.com (2018)
8. Buterin, V.: EIP 210: blockhash refactoring. Technical report (Feb 2017)
9. Buschbeck, S.: Nimiq developer reference. https://nimiq-network.github.io/developer-reference/chapters/block.html#interlink (2018)
10. Kiayias, A., Miller, A., Zindros, D.: Non-Interactive Proofs of Proof-of-Work (2017)
11. Zamyatin, A., Stifter, N., Judmayer, A., Schindler, P., Weippl, E., Knottenbelt, W., Zamyatin, A.: A wild velvet fork appears! inclusive blockchain protocol changes in practice. In: International Conference on Financial Cryptography and Data Security, Springer (2018)
12. Karantias, K.: Enabling NIPoPoW applications on bitcoin cash. Master's thesis (2019). University of Ioannina, Ioannina, Greece
13. Kiayias, A., Zindros, D.: Proof-of-work sidechains. In: International Conference on Financial Cryptography and Data Security, Springer (2019)
14. Kiayias, A., Gaži, P., Zindros, D.: Proof-of-stake sidechains. In: IEEE Symposium on Security and Privacy, IEEE (2019)
15. Dwork, C., Naor, M.: Pricing via processing or combatting junk mail. In: Brickell, E.F. (ed.) CRYPTO'92. LNCS, vol. 740, pp. 139–147. Springer, Heidelberg (1993)
16. Bonneau, J., Miller, A., Clark, J., Narayanan, A., Kroll, J.A., Felten, E.W.: SoK: research perspectives and challenges for bitcoin and cryptocurrencies. In: 2015 IEEE Symposium on Security and Privacy, IEEE Computer Society Press, pp. 104–121 (2015)
17. Bellare, M., Rogaway, P.: Random oracles are practical: a paradigm for designing efficient protocols. In: Ashby, V. (ed.) ACM CCS 93, pp. 62–73. ACM Press (1993)
18. Papadakis, T.: Skip lists and probabilistic analysis of algorithms. Ph.D. thesis (1993). University of Waterloo
19. Pugh, W.: Skip lists: a probabilistic alternative to balanced trees. In: Workshop on Algorithms and Data Structures, Springer, pp. 437–449 (1989)
20. Merkle, R.C.: A digital signature based on a conventional encryption function. In: Conference on the Theory and Application of Cryptographic Techniques, Springer, pp. 369–378 (1987)
21. Bartoletti, M., Pompianu, L.: An analysis of Bitcoin OP_RETURN metadata. In: International Conference on Financial Cryptography and Data Security, Springer, pp. 218–230 (2017)

22. Buterin, V. et al.: A next-generation smart contract and decentralized application platform. White Paper (2014)
23. Wood, G.: Ethereum: a secure decentralised generalised transaction ledger. Ethereum Project Yellow Paper **151**, 1–32 (2014)
24. Christoglou, G.: Enabling crosschain transactions using NIPoPoWs. Master's thesis (2018). Imperial College London

On Comparing the Influences of Exogenous Information on Bitcoin Prices and Stock Index Values

Luis Montesdeoca and Mahesan Niranjan

Abstract We consider time series analysis on cryptocurrencies such as Bitcoin. The traded values of any financial instrument could be seen as being influenced by market forces as well as underlying fundamentals relating to the performance of the asset. Bitcoin is somewhat different in this respect because there isn't an underlying asset upon which its value may depend on. Here, by constructing a simple linear time series model, and by attempting to explain the variation in the residual signal by means of macroeconomic and currency exchange variables, we illustrate that the influencing variables are vastly different for cryptocurrencies from a stock indices (S&P 500) in both timescales analysed (daily and monthly values). We use a sequential estimation scheme (Kalman filter) to estimate the autoregressive model and a sparsity inducing linear regression with lags (LagLasso) to select relevant subsets of influencing variables to compare.

Keywords Bitcoin · Kalman filter · LagLasso

1 Introduction

Financial time series analysis is a topic of much interest, attracting the use of various models that has gone beyond simply making accurate predictions to understanding causal relationships [4, 11]. Recent studies have focused on cryptocurrencies which are financial time series that have increased in popularity with the advancement of Blockchain. They work on a decentralized peer-to-peer network platform [14],

L. Montesdeoca (✉) · M. Niranjan
Electronics and Computer Science, University of Southampton, Southampton, UK
e-mail: ljm1e14@ecs.soton.ac.uk

M. Niranjan
e-mail: mn@ecs.soton.ac.uk

P. Pardalos et al. (eds.), *Mathematical Research for Blockchain Economy*, Springer Proceedings in Business and Economics,
https://doi.org/10.1007/978-3-030-37110-4_7

different from conventional currency systems that are regulated by central banks [2]. In conventional economy, currencies exchange rates are determined by underlying forces of demand and supply, consumer behaviour, consumers expectation of future price changes and in macroeconomic factors. In contrast none of the economic indicators drive the transaction volume of Bitcoin [20]. Due to this fact, in [6] is provided an interesting evidence that Bitcoin can be used as a strong hedge against stock market indices such as the Euro-Index for monthly returns. Also forecasting cryptocurrencies has been increasing interest for researchers that have applied multivariate analysis techniques [3, 5, 8], machine learning models [1, 9, 13] and deep learning techniques [7, 10].

Financial time series arise from complex interactions in the financial markets that include speculation, performance of companies whose shares are traded, macroeconomic variables and policy announcements. This complex interaction leads to an equilibrium between information that might be contained in the past values of a time series and new information arriving from exogenous sources. A particular way of integrating time series analysis from a statistical perspective with exogenous information is the work of Mahler [12] referred to as the LagLasso method which we pursue in this work. Here, a simple autoregressive time series model is applied to financial data such as the S&P 500 index values. The residual error in modelling is explained via a sparsity inducing regression which expresses the residual as a weighted sum of several macroeconomic variables. The sparsity constraint has the effect of selecting a small number of variables and their corresponding lags as explanatory variables of the residual. The approach is discussed in detail in the Sect. 2 of this paper.

In this paper, we compare the application of Mahler's LagLasso method on time series of cryptocurrencies and stock Indices. The hypothesis would be that the differing nature of these two market instruments would mean that the LagLasso method should find very different macroeconomic variables as explanatory variables. The difference arises from the fact that the stock index is driven by its constituent assets whose values are determined largely by their performance such as profitability, market capture and dividend payments. On the other hand, cryptocurrencies do not have any such underlying fundamentals that influence them. Their values would be dominated largely by speculative behaviour of investors and traders.

The remainder of the paper is organised as follows. We describe the LagLasso model in Sect. 2.1 and other implementation details in Sect. 2.2, followed by description of the datasets used in Sect. 3. In Sect. 4, we present results of simulations and end with some general conclusions in Sect. 4.

2 Methods

We use an autoregressive model (AR) and fit it using a Kalman filter. The AR model is defined by $\hat{x}_t = \sum_{j=1}^{p} \theta_j x_{t-j}$,

Which in vector notation is writen as $\hat{\boldsymbol{x}}_t = \boldsymbol{\theta}^T \boldsymbol{y}_t$. Where $\boldsymbol{y}_t$ contains the past values of the time series and $\boldsymbol{\theta}$ are the regression coefficients.

2.1 LagLasso

The LagLasso model, explaining the residual r by exogenous variables is also a linear model, given by Eq. 1

$$r_t = \sum_{k=1}^{K} \sum_{j=1}^{J} w_{jk} u_j(t-k), \tag{1}$$

where w_{jk} are the unknown parameters to be estimated, $u_j(t)$; $j = 1, \ldots, J$ are the variables and $k = 1, \ldots, K$ are the corresponding lags.

Selecting a subset of the variables by searching all possible combinations is a hard problem and is usual to approximate this process by including an l_1 penalty to the regression [18, 19, 21]. The resulting minimisation problem is:

$$\min_{w} \|\boldsymbol{r} - \boldsymbol{X}\boldsymbol{w}\|_2^2 + \gamma\|\boldsymbol{w}\|_1, \tag{2}$$

where $\boldsymbol{r}$ is the residual signal being modelled, $\boldsymbol{X}$ the design matrix consisting of the exogenous time series $u_j(t)$; $j = 1, \ldots, J$, γ is a hyperparemeter controlling the amount of sparsity achievable and $\boldsymbol{w}$ are the unknown parameters to be estimated.

Algorithm 1 describes the computations in pseudo-code format.

Algorithm 1 LagLasso Algorithm

```
1: Get the residuals r from applying Kalman filtering and set as target
2: Choose the number of lags {k ∈ [1, 3]}
3: Build M with the independent financial variables that you want to consider
4: Transform M to X which include k lags
5: while not all γ values are entered do
6:    Apply Lasso: min{ ||r − Xw||_2^2 + γ||w||_1 }
7:    Get non-zero values from the weight vector w
8: end while
9: Choose the most influential variables of each lag by the tuned γ selected
```

2.2 Kalman Filtering

The Kalman filter equations are given by:

$$\begin{aligned}
\boldsymbol{\theta}_{t|t-1} &= \boldsymbol{\theta}_{t-1|t-1} \\
\mathbf{P}_{t|t-1} &= \mathbf{P}_{t-1|t-1} + \boldsymbol{Q} \\
r_t &= f_t - \mathbf{y}_t^T \boldsymbol{\theta}_{t|t-1} \\
\mathbf{K}_t &= \mathbf{P}_{t|t-1}\mathbf{y}_t(\mathbf{y}_t^T \mathbf{P}_{t|t-1}\mathbf{y}_t + \boldsymbol{R})^{-1} \\
\boldsymbol{\theta}_{t|t} &= \boldsymbol{\theta}_{t|t-1} + \mathbf{K}_t r_t \\
\mathbf{P}_{t|t} &= \left(\mathbf{I} - \mathbf{K}_t \mathbf{y}_t^T\right)\mathbf{P}_{t|t-1}.
\end{aligned} \tag{3}$$

Here, r_t is the signal being modelled, $\boldsymbol{y}_t$ the vector of past values (with dimensions equal to the order of the model). $\boldsymbol{\theta}_{t|t-1}$ and $\boldsymbol{P}_{t|t-1}$ are predictions of the parameters and error covariances in them respectively. The parameters $\boldsymbol{R}$ and $\boldsymbol{Q}$ are the observation and evolution covariance matrices that tune the entire Kalman filter. The Kalman algorithm goes through a series of prediction - correction steps: predicting $\boldsymbol{\theta}_{t|t-1}$ and $\boldsymbol{P}_{t|t-1}$ from $\boldsymbol{\theta}_{t-1|t-1}$ and $\boldsymbol{P}_{t-1|t-1}$ (the first two equations). It then makes a prediction of the signal and calculates its residual error r_t. This residual and a term know as the *Kalman gain* are used in the posterior updates going from $\boldsymbol{\theta}_{t|t-1}$ and $\boldsymbol{P}_{t|t-1}$ to $\boldsymbol{\theta}_{t|t}$ and $\boldsymbol{P}_{t|t}$ respectively.

3 Data

We have collected Blockchain data and financial information from August 2011 to February 2019. It consists of 34 exogenous variables information that contains macroeconomic variables of USA market, others countries stock market index, currencies exchange rate and Blockchain information related to Bitcoin. Table 1 shows this data that was acquired from Thomson Reuters Datastream Platform at University of Southampton with daily and monthly values. In addition to Blockchain information and USA market, we consider the commodities: oil, gas and gold price due some works analyze the effect of those in stocks and currencies [15, 17].

Table 1 Cryptocurrency information and financial data used in this work found from Thomson Reuters datastream platform at University of Southampton

CPTRA: Cost/Trans.	ETRAV: Estimated Trans. Vol.	TOUTV: BTC Total Output Vol.
MIREV—BTC Miners Revenue	NADDU: Num. Unique Addresses	NTRBL: Num. Trans/Block
NTREP: Trans. Exc. Popular Addr.	TRFEE: BTC Total Trans. Fees	TRVOU: Exch. Trade Vol.
NTRAT-Total Num Transactions	HRATE: BTC Hash Rate	MKTCP—BTC Market Cap.
Virtual Crypto Technologies	USA Amount Market	Government Budget
Equity Risk Premium	Oil WTI	Gas
Gold	Market Issues	Global Investors
Trade Balance	10Y Bonds	EUR/USD
GBP/USD	Yen/USD	Yuan/USD
Nikkie 225	DAX 30	FTSE100
Policy Uncertainity	Personal Incomes	Infl-LKD 10Y Bld
Production		

4 Results

Figure 1 shows the time series being modelled and the corresponding residuals when an autoregressive model of order three is applied.

We note a clear reduction in the variance of the residual which also appears to be zero mean, as expected. Figure 2 shows how the number of macroeconomic variables getting non-zero values at increasing levels of the regularization parameter γ for the daily timescale (similar effect was found for the monthly values). There is a monotonic decrease in the number of parameters, again as expected. Inspecting this graph we chose 23 variables for the Bitcoin data and 28 variables for the S&P 500 data (where there is a flat region in the graph). This selection is indeed a matter of convenience and in a practical situation of applying such a technique some higher level consideration needs to be brought in. Here, it suffices to say that the prominent explanatory variables is what we seek.

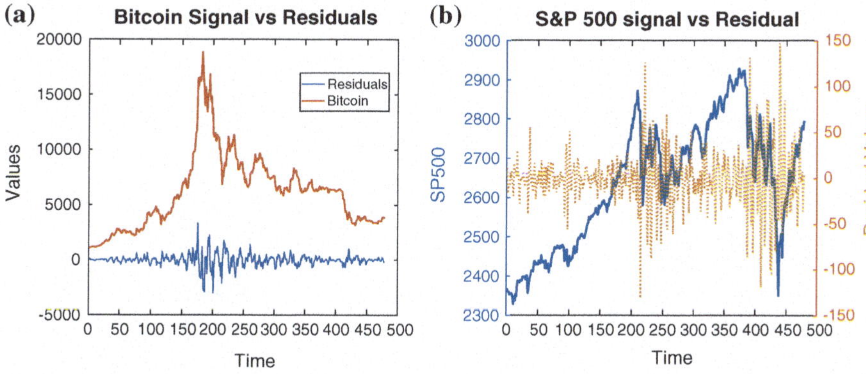

Fig. 1 The target time series (Bitcoin values (**a**) and S&P 500 index (**b**)) and the corresponding residuals after Kalman filtering. Initial X axis values are from the time that it converged for better illustration

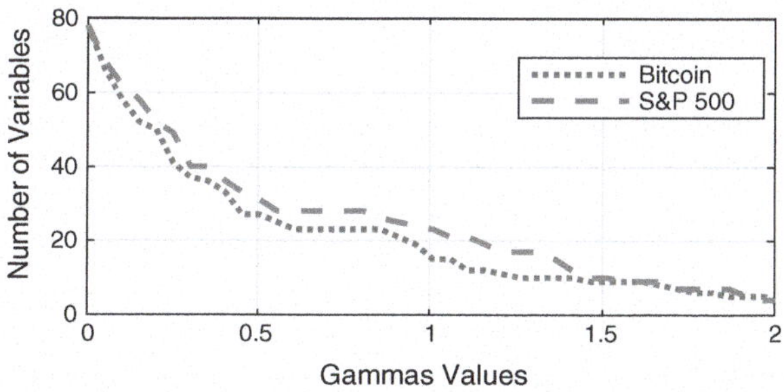

Fig. 2 The number of variables (and lags) selected as the amount of regularisation is increased. In general, there will be a monotonic decrease in the number of non zero coefficients, and we look for a plateau or knee in the graph to select a covenient subset

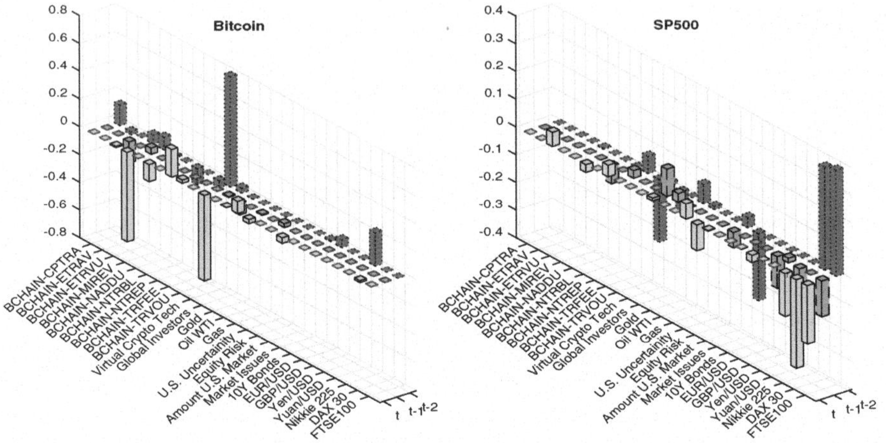

Fig. 3 Influence of the different macroeconomic variables on the cryptocurrency and stock index time series with daily values, separated by the three lags used. The financial information that affect Bitcoin do not have such underlaying fundamental that influence S&P 500

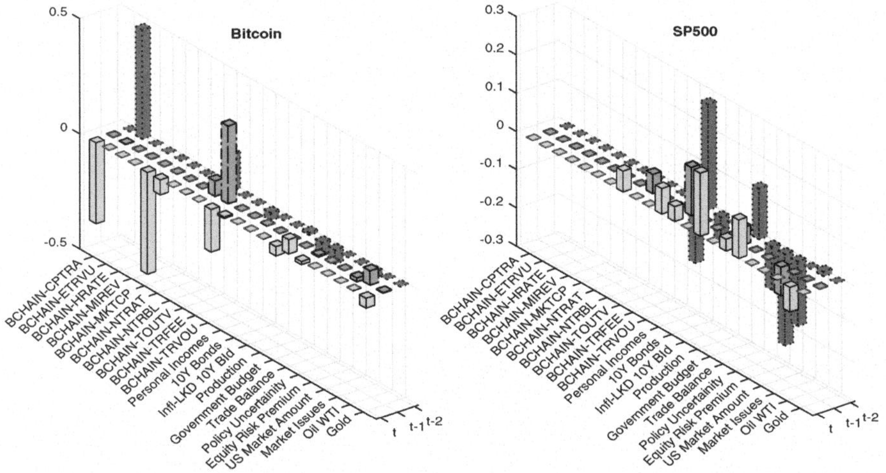

Fig. 4 Influence of monthly macroeconomic variables values on the cryptocurrency and stock index data. There is clear difference as Fig. 3 in the type of variables that influence Bitcoin and S&P 500

Figure 3 shows the influence of the variables as given by the corresponding weights of the regression solution, separated by the three lags used on daily values, similar to Fig. 4 where monthly values were analysed. Finally we also found clear difference in the type of variables that influence Ethereum (another popular cryptocurrency) and Dow Jones, which support our analysis of cryptocurrencies do not have any such underlying fundamentals that influence them as the stock market Indices.

5 Conclusion

This work illustrate a fundamental difference between the traded values of cryptocurrencies (such as Bitcoin) and other financial assets (such as stock indices), in that expalanatory exogenous variables of relevance are not the same between them in the two timescales analysed (daily and monthly values). We postulate that this is because cryptocurrency values respond primarily to trader sentiments and objectives. Unlike stock indices, cryptocurrencies do not have any underlying assets of economic performance to modulate their values. This is shown by the technique of a sparsity inducing regularized linear regression modelling residual signals of an autoregressive process applied to Bitcoin and S&P 500 data. Our current work is focused on similar analysis with structured matrix factorization methods, again with exogenous variables as additional inputs starting from the work in [16].

Acknowledgements Mahesan Niranjan acknowledges support from the project Joining the Dots: From Data to Insight (EP/N014189/1), funded by the Engineering and Physical Sciences Research Council, UK.

References

1. Almeida, J., Tata, S., Moser, A., Smit, V.: Bitcoin prediction using Artificial Neural Networks. Neural Networks, pp. 1–12 (2015)
2. Antonopoulos, A.M.: Mastering Bitcoin: unlocking Digital Cryptocurrencies. O'Reilly Media Inc., (2014)
3. Bakar, N.A., Rosbi, S.: Autoregressive integrated moving average (ARIMA) model for forecasting cryptocurrency exchange rate in high volatility environment: a new insight of Bitcoin transaction. Int. J. Adv. Eng. Res. Sci. **4**(11) (2017)
4. Billio, M., Getmansky, M., Lo, A.W., Pelizzon, L.: Econometric measures of systemic risk in the finance and insurance sectors. Technical Report, National Bureau of Economic Research (2010)
5. Catania, L., Grassi, S., Ravazzolo, F.: Forecasting cryptocurrencies under model and parameter instability. Int. J. Forecast. **35**(2), 485–501 (2019)
6. Chan, W.H., Le, M., Wu, Y.W.: Holding Bitcoin longer: the dynamic hedging abilities of Bitcoin. Q. Rev. Econ. Financ. **71**, 107–113 (2019)
7. Guo, T., Bifet, A., Antulov-Fantulin, N.: Bitcoin volatility forecasting with a glimpse into buy and sell orders. In: 2018 IEEE international conference on data mining (ICDM), pp. 989–994. IEEE (2018)
8. Hotz-Behofsits, C., Huber, F., Zörner, T.O.: Predicting crypto-currencies using sparse non-Gaussian state space models. J. Forecast. **37**(6), 627–640 (2018)
9. Jang, H., Lee, J.: An empirical study on modeling and prediction of Bitcoin prices with bayesian neural networks based on Blockchain information. IEEE Access **6**, 5427–5437 (2018)
10. Karakoyun, E.Ş., Çıbıkdiken, A.O.: Comparison of ARIMA time series model and LSTM deep learning algorithm for Bitcoin price forecasting. Econ. Manag. Mark. (MAC-EMM 2018) 171 (2018)
11. Kwan, A.C., Sim, A.B., Cotsomitis, J.A.: The causal relationships between equity indices on world exchanges. Appl. Econ. **27**(1), 33–37 (1995)

12. Mahler, N.: Modeling the S&P 500 index using the Kalman filter and the LagLasso. In: IEEE International Workshop on Machine Learning for Signal Processing, 2009. MLSP 2009, pp. 1–6. IEEE (2009)
13. McNally, S., Roche, J., Caton, S.: Predicting the price of Bitcoin using machine learning. In: 2018 26th Euromicro International Conference on Parallel, Distributed and Network-Based Processing (PDP), pp. 339–343. IEEE (2018)
14. Narayanan, A., Bonneau, J., Felten, E., Miller, A., Goldfeder, S.: Bitcoin and Cryptocurrency Technologies: a Comprehensive Introduction. Princeton University Press (2016)
15. Pukthuanthong, K., Roll, R.: Gold and the dollar (and the euro, pound, and yen). J. Bank. Financ. **35**(8), 2070–2083 (2011)
16. Squires, S., Montesdeoca, L., Prügel-Bennett, A., Niranjan, M.: Non-negative matrix factorization with exogenous inputs for modeling financial data. In: International Conference on Neural Information Processing, pp. 873–881. Springer (2017)
17. Sujit, K., Kumar, B.R.: Study on dynamic relationship among gold price, oil price, exchange rate and stock market returns. Int. J. Appl. Bus. Econ. Res. **9**(2), 145–165 (2011)
18. Takeda, A., Niranjan, M., Gotoh, J., Kawahara, Y.: Simultaneous pursuit of out-of-sample performance and sparsity in index tracking portfolios. Comput. Manag. Sci. **10**(1), 21–49 (2013)
19. Tibshirani, R.: Regression shrinkage and selection via the Lasso. J. R. Stat. Soc. Ser. B (Methodological) 267–288 (1996)
20. Vaddepalli, S., Antoney, L.: Are economic factors driving Bitcoin transactions? an analysis of select economies. J. Emerg. Issues Econ. Financ. Bank. (JEIEFB) **6**(2), 2215–2227 (2017)
21. Weston, J., Elisseeff, A., Schölkopf, B., Tipping, M.: Use of the zero-norm with linear models and kernel methods. J. Mach. Learn. Res. **3**(Mar), 1439–1461 (2003)

Performance of Tip Selection Schemes in DAG Blockchains

Richard Gardner, Philipp Reinecke and Katinka Wolter

Abstract In this paper we investigate the impact of transaction validation of two tip selection mechanisms in DAG blockchains such as the tangle of IOTA on the performance of the consensus mechanism. The tip selection algorithm determines which prior transactions are validated by a transaction. With validating a tip a transaction is appended to the tangle. We present TangleSim, our simulator based on OMNeT++, which allows to evaluate the transaction validation time and the time a transaction will spend as a tip in DAG blockchains. We find that the weighted random walk selection can achieve a lower transaction validation time than the random tip selection algorithm in many cases.

Keywords Tip selection · Simulation · Tangle · IOTA

1 Introduction

Blockchain technologies have gained much public attention recently. However, they still polarise the public into those who are convinced by their potential to provide a trust-worthy distributed infrastructure and those who reject them because of the most commonly applied consensus mechanism, proof of work (PoW), also known as mining. The most notable application of blockchain technologies are cryptocurrencies, and among them, Bitcoin [1] and Ethereum [2] are still the most popular. At the same time, Bitcoin is heavily criticised because of its high energy usage and low transaction rate. Bitcoin's transaction throughput is not more than 13 Tx/s [3],

---

The work was done while the author was at Cardiff University.

R. Gardner · P. Reinecke
Cardiff University, Cardiff, UK
e-mail: ReineckeP@cardiff.ac.uk

K. Wolter (✉)
Free University Berlin, Berlin, Germany
e-mail: katinka.wolter@fu-berlin.de

P. Pardalos et al. (eds.), *Mathematical Research for Blockchain Economy*,
Springer Proceedings in Business and Economics,
https://doi.org/10.1007/978-3-030-37110-4_8

which is not competitive compared to commercial systems such as e.g. PayPal (295 Tx/s) [4].[1]

The cryptocurrency IOTA [5] aims to create a light-weight blockchain with short transaction confirmation times (and hence high transaction throughput). IOTA uses PoW and transaction validation to create its blockchain, like Bitcoin. Instead of a chain of blocks containing an arbitrary number of transactions, IOTA transactions are stored in a Directed Acyclic Graph (DAG), referred to as the Tangle. Transactions are atomic in the tangle—meaning one block is one transaction, and each transaction is connected via edges to other transactions, where an edge means a transaction has approved another. To issue a transaction and store it permanently the issuer must approve two other transactions, where a transaction that has not yet been approved is a *tip* [6]. The issuer uses a tip selection algorithm to choose which tips to approve and computes the proof of work ensuring their validity. Ideally, all transactions would be approved within a very short time, storing them permanently on the IOTA tangle. In practice, however, there are many transactions with more than one or two approvers and others that must wait very long to be approved.

In this paper we study the performance of two tip-selection methods, viz. Uniform random tip selection(URTS) and the Weighted random walk (Walk). Both strategies have been studied before [6] and are explained and illustrated in [7]. Prior work mostly serves the purpose of explaining the functioning of the tip selection algorithms in a visual simulation. Our work aims at exploring which tip selection strategy can achieve the higher transaction throughput and arrives at shorter time for a transaction to obtain the necessary approvals. URTS chooses a tip completely at random, placing no higher value in choice from one tip to the next. The Walk selection method traverses the tangle from transaction to transaction from a determined distance back until a tip is reached. This tip is then used as the attachment site.

As discussed in [5], these methods differ in their impact on the tangle: URTS has potentially higher performance, but it does not discourage lazy or malicious entities from approving either the same transaction repeatedly, or forming a malicious sub-tangle that could enable a double spend. The purpose of the Walk selection method is to provide a slight bias towards higher weighted transactions i.e. those that have been approved or indirectly approved by more transactions, this creates a structure that is much harder to compromise and is essential to a tangle cryptocurrency. Our simulations do not confirm the superiority of URTS in throughput and latency.

We present the discrete-event simulation TangleSim.[2] which we use to evaluate the number of tips seen by an issued transaction, the time as a tip and the time until approval of the transaction under different load.

This paper makes the following contributions:

- we simulate a tangle with non-uniform population (slow nodes and fast nodes), which achieve slow or fast transaction approval

[1] PayPal, during second quarter of 2018 processed 2,327,000,000 transactions which equates to an average of 294.93 Tx/s.

[2] Available on Github at https://github.com/richardg93/TangleSim.

– we compute three different metrics, the number of tips seen by an issued transaction, the time as a tip and the time until approval of the transaction under increasing load
– we publish the code of TangleSim for further use by the interested reader.

The remainder of the paper is structured as follows: In Sect. 2 we provide necessary background knowledge on the mechanics of a blockchain and on the details of IOTA. In Sect. 3 we discuss metrics for evaluating the performance of tip-selection methods. We then introduce our simulation model (Sect. 4), before presenting the results in Sect. 5.

2 Background

The typical purpose of a blockchain is to enable purely peer to peer payments, without relying on trusted third parties, as described in [1]. More generally, the purpose can be summarised as providing a distributed transaction network without any centralised entity having authority over transactions. In the Bitcoin [1] network transactions are broadcast to nodes which collect a set of transactions and then compete to solve a complex puzzle. On completion the solver (*miner*) may append a new block to the chain. This new block is broadcast to all other nodes who verify the authenticity of the completed puzzle. They show their acceptance of the block by starting to work on the next block.

This complex puzzle is called proof of work (PoW). It allows participants in the network to be confident that the ledger of transactions is valid as long as at least 50% [1][3] of the computing power in the network is contributing honestly. In the Bitcoin and Ethereum [2] protocols there are two roles: *miners* and *transactors*. These networks provide an incentive to miners by rewarding the node that solves the puzzle first with a significant amount of Bitcoins or Ether. Second, transactors must pay a transaction fee (included in the transaction itself) which goes to the miner who finishes the block said transaction is included in. The mining process in proof of work is tuned by the *difficulty*, which determines how long it takes for a block to be mined, or for the miners to meet the corresponding *target*. Another important parameter is the block size, i.e. how many transactions can be fit into any given block in the chain.

2.1 IOTA

The IOTA token [5] is similar to many cryptocurrencies in that consensus is achieved through proof of work.[4] However, this is where the similarities end in the actual

[3]It has been shown that given certain conditions and strategies this proportion can decrease to 30% [8].

[4]Other currencies such as Nano [9] and Dash [10] use proof of stake.

implementation. Instead of a chain of blocks containing an arbitrary number of transactions, IOTA transactions are stored in the *Tangle*, which is a Directed Acyclic Graph. Transactions are atomic in the tangle, i.e. one block is one transaction. Transactions are connected via directed edges, where an edge means that the transaction at its origin has approved the transaction at its destination. When a transactor wishes to issue a transaction the issuer must approve two others. An unapproved transaction is known as a *tip*, The issuer uses a tip selection algorithm to choose which tips to approve and then computes the proof of work ensuring their validity.

Unlike in Bitcoin, IOTA transactions are an atomic part of the data structure. This means that the maximum size of any single state change is exactly one transaction, and there can be no state change of a different size, unlike in Bitcoin where a state change can range from one to the maximum number of transactions in a block, every 10 min.[5] There are no fixed time intervals, so a transaction can be attached and made available for approval as and when it is needed, with no waiting for the next block. This avoids the block size and interval problems of blockchain by using a data structure with finer granularity.

2.2 Tip Selection Methods

When issuing a transaction the transactor must select two transactions to approve from all available transactions. The tips should be selected in a way that supports the health of the tangle. As discussed in [7] a good tip selection algorithm should make sure that no transactions are left behind unconfirmed and the number of direct and indirect confirmations of each transaction should grow over time. The uniform random tip selection as well as the weighted random walk tip selection studied in this paper both mostly avoid these problems.

Uniform random tip selection (URTS) The uniform random tip selection algorithm picks a tip at random uniformly distributed from the transactor's local copy of the global tip list.

This is illustrated in Fig. 1: The white squares represent transactions with one or more approvers. The black squares are unappproved transactions (tips). No new transaction has decided until this moment to attach themselves at tips A, B and C.

Block D, a newly issued transaction, can see that it has a choice of the three tips. The simplest way it can do this is to pick from these three at random, without any bias at all. Each transaction selects two tips to approve and once chosen will attach itself to those two. Here, newly issued transaction D, chooses the simple approach and selects at random A and B, itself becoming a tip.

Weighted random walk tip selection (Walk) The weighted random walk tip selection is a more complex method. It traverses the graph from an older transaction

[5]The network choosing to discard a sub-chain for a longer sub-chain could be considered a state change as well. In contrast, in the Tangle transactions attached are never discarded, but can be left behind when other transactions choose not to approve them.

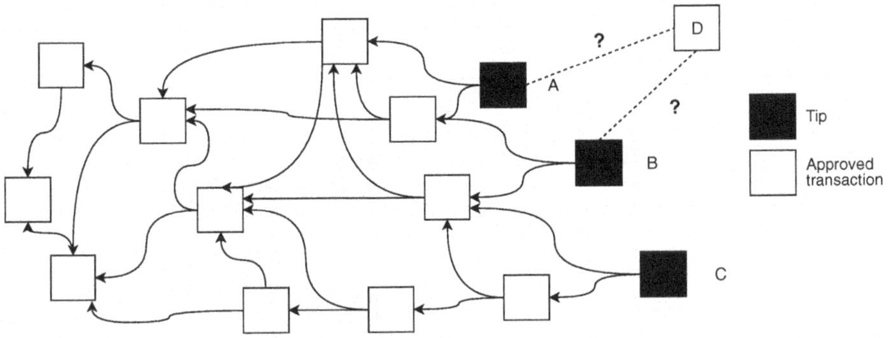

Fig. 1 Uniform random tip selection (URTS)

towards the newer transactions until a tip is found. Decisions at each step are made as follows: With probability α, the highest-weighted transaction is chosen, and with probability $1 - \alpha$ it picks at random from the available transactions. It is not clear from the literature whether IOTA chooses randomly from all or only amongst the ones that are not the highest-weight one. Our simulation chooses amongst all the available transactions, including the highest weighted transaction.

The starting point is chosen by a uniform random backtrack to a depth of $20 * \lambda$, where λ is the rate of transactions per second (see Fig. 2). This decision is based on the finding in [6] that a depth higher than 20λ has no additional effect on the number of tips at any given time in the network. The parameter λ is the flow rate of transactions in the tangle in [6].

In the simulation model described in this paper, Walk tip selection is implemented via the use of a walker particle released by the newly issued transaction.[6] This walker selects a single new tip by the following process, composed of 2 phases:

1. A backtrack into the tangle, as shown in Fig. 2: The algorithm chooses a tip at random using the URTS algorithm. It then moves from edge to edge away from the tip until a distance (in terms of edges traversed) of $20 * \lambda$ (i.e. 20 for $\lambda = 1$ and 60 for $\lambda = 10$) has been reached or the genesis transaction is found (whichever happens first). The choice of which edge to follow is determined completely at random with no bias towards any transaction in particular. The transaction on which the walk stops serves as the starting point for the walk forward towards the tips.
2. The walk towards the selected tip uses the cumulative weight of a transaction, as introduced in [7]: An unapproved transaction has weight 1 (e.g. the tips in Fig. 3. Transaction X has a weight of 2 as it has one direct approver, while transaction Y has a weight of 6, as there are 3 other transactions that approve those transactions that have themselves approved transaction Y. Transaction Z is the genesis transaction, this means that all transactions in the tangle approve it directly or

[6]The IOTA team place their walkers at regular intervals, it is not clear exactly what mechanism they use to achieve this. TangleSim uses the backtrack method per walker.

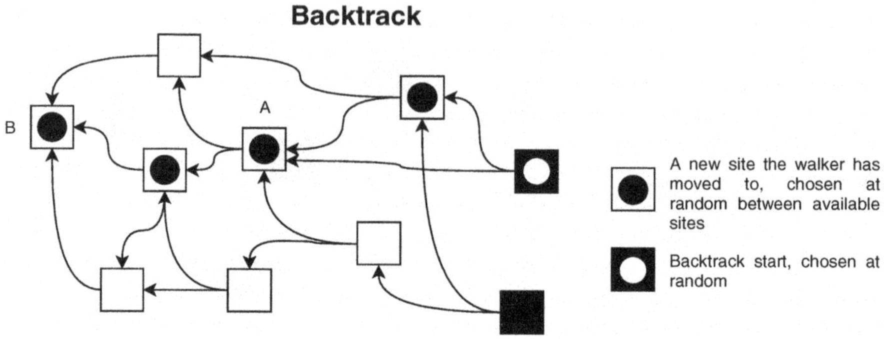

Fig. 2 Backtracking to find the starting point for the random walk

indirectly. This gives transaction Z a cumulative weight of the total transactions in the tangle, including itself—in this case 10.

For the walk forwards towards the tips, the walker's decision at every site is no longer completely at random. Our walker is equipped with a parameter α. This value is used to determine the priority of our choice between the available sites at each transaction the walker visits, and is a number between 0 and 1. In every step, with probability α the walker chooses the transaction with the highest weight. With probability $1 - \alpha$ it picks one of the available sites using a discrete uniform distribution (in exactly the same way as the backtrack). This process is repeated until our walker reaches a tip (i.e. a transaction with no or one approver(s)), which is then passed to the newly issued transaction to attach to. The process is repeated for the second tip. Figure 4 illustrates this: In move A the walker chose to move to the highest weighted transaction. In move B, our walker yet again moves to the highest weighted site available. The same thing happens in move C. In move D, the walker picks at random from the available sites. In this case, the random choice determines that it must move to a site which happens to be a tip—thus ending the walkers search for a tip.

2.3 *Differences Between IOTA, Other Work and TangleSim*

Simulations for the tangle were discussed in [6, 7]. These simulations are designed to illustrate the dynamics and algorithms of IOTA while extracting meaningful insight about the tangle structure. In contrast, our simulation is driven by the wish to assess the performance of the tip-selection algorithms in a scenario closer to real life. Perhaps the biggest difference in the model is our implementation of transactors in the tangle, in that we assign a specific time for an individual to compute their proof of work to validate a transaction, drawn from a distribution so as to simulate a network of unique individuals (cf. Fig. 4).

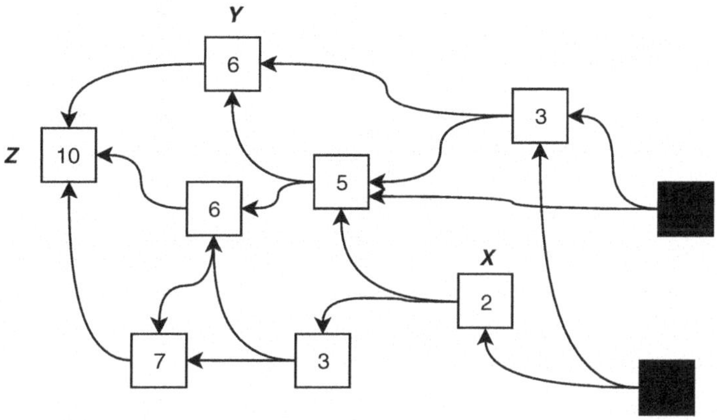

Fig. 3 Weight computation for the random walk tip selection

TangleSim can use a variable number of walkers, while in IOTA [6] the number of walkers is equal to $3k + 4$, where k is the number of tips a newly issued transaction must approve. The first walkers to find a tip to satisfy k are chosen, the rest are discarded. The reasoning here is to combat potential parasitic chain attacks and to ensure that the same tip is never chosen twice. We chose one walker in most experiments for two reasons: Using Walk tip selection requires that the weights of transactions a walker encounters be calculated. This is computationally expensive, and calculated by recursively traversing the tangle in this model. Secondly, by releasing many walkers and accepting those first to return, we are showing a bias towards tips with a lower weight, whether this has been considered by the IOTA team remains unclear. We submit that choosing at random from among all the tips chosen by the walkers would eliminate any bias towards those walkers that make unlikely choices to unattractive tips (such as those attaching a parasitic chain).

In a real tangle network nodes would know about other transactions from broadcasts of other nodes. It is unclear how in the IOTA simulation model an issuer's view of the tips takes this into account. The fact has been merely stated in [6]. In TangleSim nodes (or transactors) are deemed to have an up to date view of the tips available to select when they start to perform tip selection, but this view is frozen while they select and subsequently compute the proof of work to validate them.

In [5] and [11], tangle performance has been studied using analytical approaches. For URTS, [5] provides estimates for the time as a tip and number of tips as a function of the transaction rate. [11] focusses on studying stable strategies for the tangle. The authors utilise Little's Law to give an estimate of the number of unapproved transactions as a function of the time after which a transaction is considered orphaned and the probability of approval within a given time period. In both papers, it is not entirely clear how the required values can be obtained. Our paper augments these studies by providing realistic simulations.

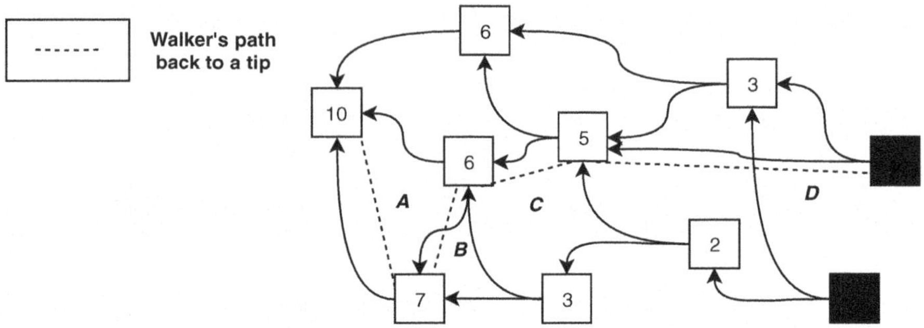

Fig. 4 Uniform random tip selection

3 Metrics

In this section we discuss the metrics which we use to assess the performance of both tip selection algorithms.

Time as a tip The time as a tip is defined as the time between the creation of a transaction and its being chosen by another transaction's tip selection algorithm: $t_{\text{taat}} := t_{\text{selected}} - t_{\text{created}}$.

Approval time The approval time measures the time until a transaction is considered unlikely to be reversed. The approval time includes the time as a tip, but is more than that. It is the time from the moment of issue of a transaction until the cumulative weight of the transaction starts to increase linearly at the global transaction-issue rate. From that point on, all transactions are (indirectly) approving the given transaction. For example, this always holds for the Genesis block, as all transactions will directly or indirectly confirm it, so its weight will increase by one for every transaction added to the system [5]. In contrast, the time as a tip is simply how long it took for the first other transaction to select this transaction using its tip selection algorithm.

4 The TangleSim Simulation Model and Its Parameters

We will now discuss our simulation model. The model uses the discrete-event simulation framework OMNeT++ [12].

4.1 Interface and Implementation

The overall structure of TangleSim is given in Fig. 5. The model provides two specialisations to the cSimpleModule base class: TxActorModule and TangleModule. These two classes provide the interface to use in the OMNeT++ environment: initialise—

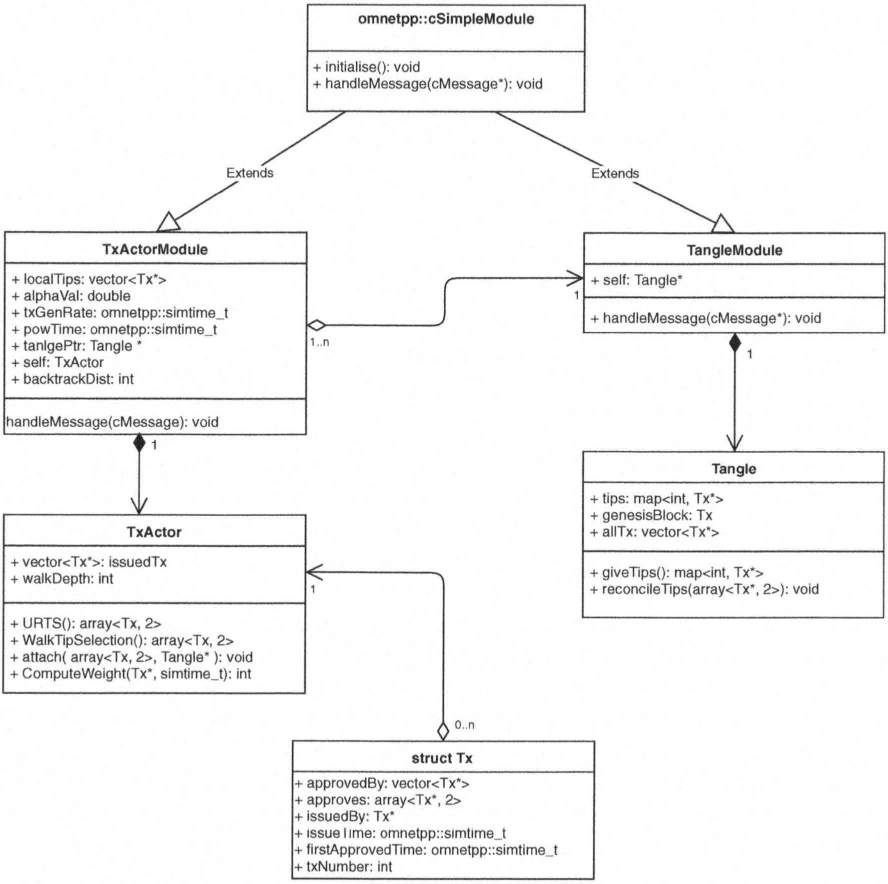

Fig. 5 UML representation of TangleSim

which sets up each module with the parameters supplied from the network description file (.ned), handleMessage—which uses the type of message received to decide a course of action, whether that be passing the message on to another module or deleting the message as it has reached the end of its lifetime.

As we can see from Fig. 5, we have two classes that include most of the implementation behind the interface that the derived classes provide. TxActorModule and TangleModule can be thought of as privately inheriting from TxActor and Tangle respectively, as these are not exposed to the OMNeT++ environment directly. Finally, Tx is essentially a data structure that represents an atomic transaction. It is the building block we use to construct a tangle from, a Tx has pointers to the transactions that directly approve it and those it directly approves. With these links between individual transactions, we are able to move from any one point in the tangle to another—which is especially useful when computing the cumulative weight of a transaction. The genesis transaction is held by the internal Tangle object, while all other transactions are created dynamically and held by the TxActors that issue them.

The Tangle object holds an up to date map containing all the transactions that currently have no approvers using the TxNumber as its key for fast access at high tip numbers. Each TxActorModule also holds its own local copy of the tip list, which it requests from the Tangle when it decides to issue a transaction.

4.2 Module Communication

A TxActorModule has a copy of the global tip list. This is important as actually approving a transaction and attaching to it involves computing some proof of work to prove that the transactions it chooses are valid. However long it takes to validate its chosen transaction, there is nothing to stop another transaction from selecting the same tip. Before either knows the other wants to approve the same one. If we imagine a simulation where all nodes select their tips from the same global list with no latency or PoW to complete, transactions would only ever have one approver—which would be the ideal situation as this means no computing power is wasted re-affirming a transaction's valididty. However, this does not reflect the real world—where nodes issuing transactions are communicating over the internet with delay, so it is important when modelling such an asynchronous system that we account for this.

As such, the event of a TxActorModule issuing a transaction is split into two distinct events: the tip selection and the attachment. The OMNeT++ environment only allows one event to happen at a time, being scheduled according to a global clock. By splitting a transaction issue into two events, we emulate a real life tangle as described above. As seen in Fig. 6 these two events are realised by the use of the cMessage send, receive and handle methods we inherit as a cSimpleModule. There are 5 types of message: NEXT_TX_TIMER, TIP_REQUEST, CURRENT_TIPS, POW_TIMER and ATTACH_CONFIRM.

NEXT_TX_TIMER is a self message a TxActorModule schedules itself to receive at the beginning of the simulation and whenever it attaches a transaction. When a NEXT_TX_TIMER is received by a TxActorModule, it sends a TIP_REQUEST message to the TangleModule, this message has no delay to reflect the fact that a node in real life would be updating its view of the tangle as it goes, the TangleModule then sends a CURRENT_TIPS message back to the TxActorModule again with no delay, this message tells the TxActorModule to copy the global tip list from the TangleModule. It can then use either URTS or Walk tip selection to select two tips. Once a TxActorModule has selected, it will send another self message POW_TIMER—to simulate the time taken to compute the validation for its chosen transactions. Once the TxActorModule receives its own POW_TIMER message, it immediately sends an ATTACH_CONFIRM message to the TangleModule. On—immediate—reception the TangleModule will compare the tips against its up to date list, and remove any that have not already been approved. Immediately after sending the ATTACH_CONFIRM message, the TxActorModule schedules another

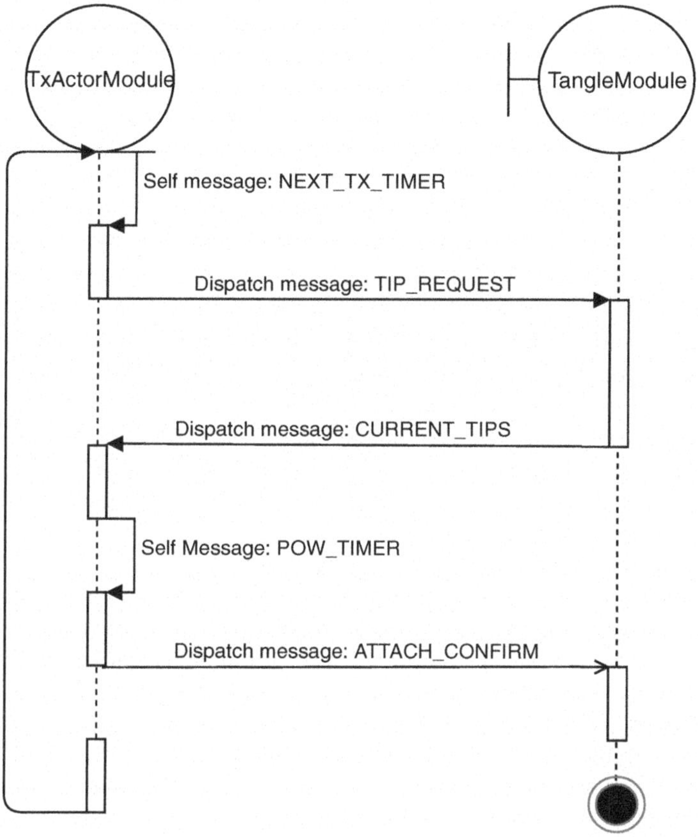

Fig. 6 Sequence diagram representing the discrete events making up a TxActor issuing a transaction

NEXT_TX_TIMER which when received will signify the start of the entire process again. All TxActors in the simulation environment will continuously repeat the above steps until a global transaction limit is reached, signalling the end of the simulation.

4.3 Parameters

As mentioned in the previous section, the TxActorModules schedule two self messages, the first to determine how often they will start the process of issuing a transaction, and the other to determine how long it takes them to compute the proof of work to validate the chosen transactions. These two times together essentially determine how often a transaction will be issued per TxActorModule. For example if we wanted one TxActorModule to issue a transaction every ten seconds, we could set the NEXT_TX_TIMER to be 9s and the POW_TIMER to 1s.

In total there are seven important parameters in any simulation: TxActorCount, TxGenrate, PowTime, TransactionLimit, AlphaValue and WalkDepth. The last two are only used if are TxActorModules are using Walk tip selection. TxActorCount is how many transactors we want in our simulation—and the global transaction issue rate is calculated from the individual transaction rate (described above) by the TxActorCount.

We ran simulations with a transaction issue rate of 1 tx/s, 3 tx/s and 10 tx/s. TxGenRate and PowTime are not the same per TxActorModule. For the simulations at 1 tx/s, 3 tx/s and 10 tx/s, the TxGenRate is drawn from an exponentially distributed random variable with a mean of 900s—this is drawn per transaction issued and used to schedule how long a NEXT_TX_TIMER message will take to return to its sender. For PowTime, this model aims to emulate a real life tangle, with many different types of transactors with different hardware to compute proof of work. To that end, PowTime is set at the start of the simulation per TxActorModule, and drawn from a truncated normal distribution (to ensure that time is not negative) with a mean of 100s and a standard deviation of 10s. The transaction rates described per simulation were engineered by keeping the TxGenRate and PowTime the same, but increasing the TxActorCount—1000, 3000 and 10000 respectively.

The Transaction Limit was originally set at 25000 transactions attached for all rates, however, we found that at 10 tx/s, the results no longer fit the pattern we had seen at lower rates as the shape of the tangle does not stabilise until a certain number of transactions has been attached. This has been observed by the IOTA research team as well. In [6] they observe that the higher the transaction rate, the longer it takes for the tangle to stabilise—or reach equilibrium, defined as when the number of tips at any given time remains fairly constant. With this in mind, the simulations at 10 tx/s were run until 50,000 and the first 25,000 transactions were excluded from the results, which neutralised the anomalous results seen. In 1 tx/s the first 5000 transactions were excluded and 10,000 for 3 tx/s. By running each simulation 30 times with from a seed sequence we were able to obtain enough data to generate coherent results.

We set $\alpha = 0.001$ in the 1,3 and 10 tx/s simulation, so a walker will have a 1/1000 chance at every site of choosing the highest weighted transaction available to it. The IOTA team have shown in [6] and [13] that significantly higher values mean too many transactions are left behind and the shape of the tangle becomes unsustainable for all transactions issued to have a decent chance of being accepted.

For the WalkDepth, the IOTA team use a different metric to describe their transaction issue, or flow rate λ (transactions issued per time unit [6]), purporting that starting a walk from a depth of $20 * \lambda$ being deep enough to have no affect on the number of tips at any given time. Yet in practise they use 100–200 $*\lambda$ to be on the safe side. In the simulation we judge the flow rate to be its average transaction rate, and so used WalkDepth values of 20, 60 and 200 for the simulations respectively.

We ran a series of simulations to determine the effect of a subset of a tangle network having a slow network connection and/or slow proof of work compute time—relative to the rest of the network. For this experiment, 4 simulations were completed. In the control run, the overall issue rate was 1 tx/s, calculated from a TxActorCount of 1000, a TxGenRate of exponential (1000 s) and a PowTime of 1s. Walk tip selection was used in all runs, with an $\alpha = 0.1$ and a WalkDepth of 20. In the three proceeding runs, 100 of the 1000 TxActorModules were given a PowTime of 1.5s, 2s and 3s respectively. Each was run until 25,000 transactions were attached (30 runs each).

5 Results

Figure 7 shows the impact of the latency on the time to approval on the left and the average time to approval on the right side.

The experiments with different latencies show that the tangle is very sensitive to such variation. Even modest differences in a small subset have a considerable impact on the approval rates for the whole network. This result illustrates why IOTA was struggling in the early days with very few users. As far as we know, stability of IOTA now is much better.

Figure 8 shows the average time as a tip on the left and the average number of tips seen by an issued transaction on the right. This result shows the superiority of the walk tip selection. The time as a tip, the time before a transaction is selected as a tip is much lower for the walk tip selection than for URTS and it decreases with

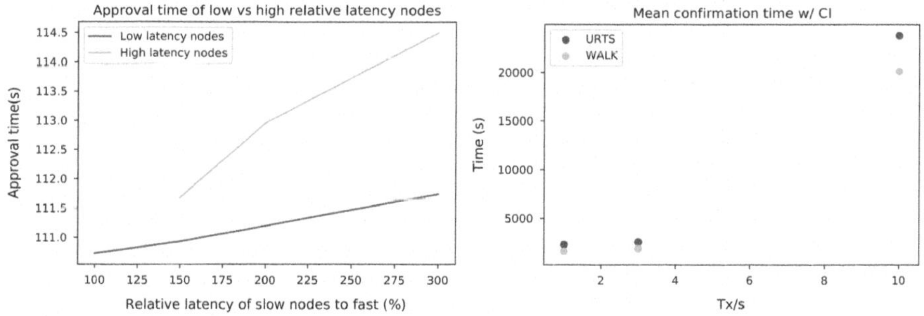

Fig. 7 Latencies

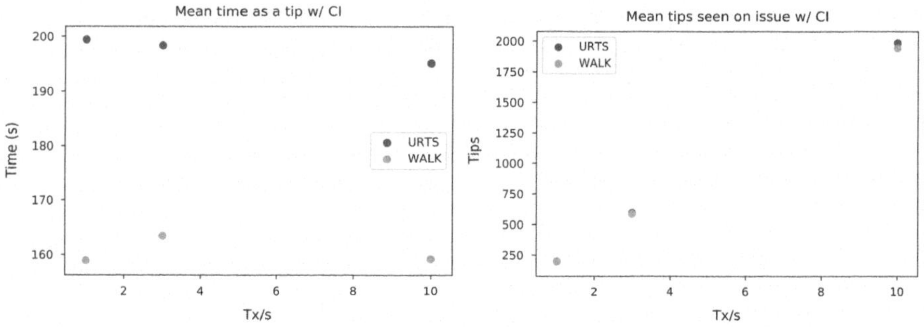

Fig. 8 Left: Avg. time from issue to first approval at different loads. Right: Avg. total tips present at the time of issue

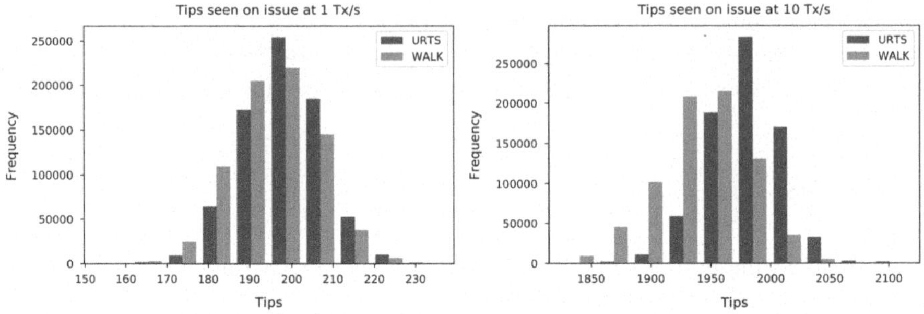

Fig. 9 Number of tips seen

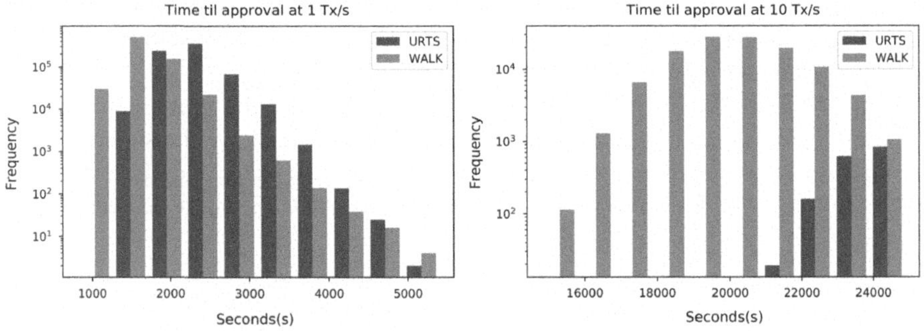

Fig. 10 Time until approval

increasing transaction load. The more transactions are submitted, the shorter is the time a transaction lives as a tip. And the walk algorithm is much better at finding good tips that reduce a transaction's time as a tip than the random selection strategy.

Consequently, as shown in Fig. 8 on the right, there are much fewer tips available with increasing transaction rate when using the walk tip selection, than when using the purely random strategy.

Even though the time as a tip can be reduced with increasing transaction rate, the time until approval increases strongly with the load as shown in Fig. 10.

Figure 9 shows the number of tips seen by an issued transaction. Using the random tip selection at high transaction rate there are many more tips than for when the walk strategy is applied.

In Fig. 10 the histogram of the time until approval under different load is shown. The walk strategy leads to more short transaction approval times even more so when the system load increases.

Please note that unapproved transactions were excluded from the graphs. In consequence the histograms for the two strategies are not based on the same number of samples.

To shed light on this behaviour Fig. 11 shows the number of tips seen per transaction from the first transaction issued to the last in our experiment series.

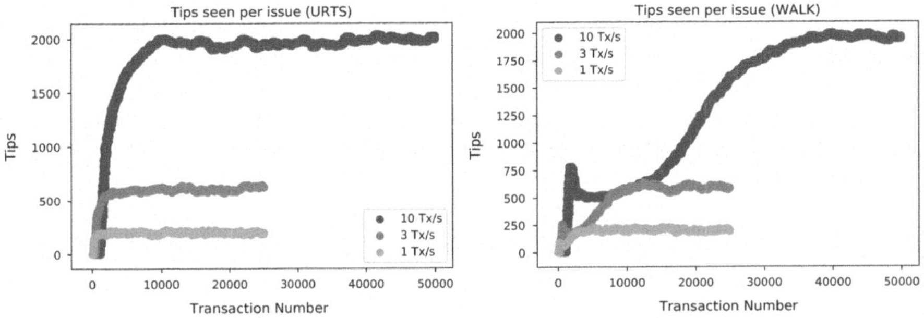

Fig. 11 Tips seen for both selection strategies

Several interesting observations can be made in the two graphs. First, the URTS selection scheme converges fast to a rather stable value which seems to be proportional to the transaction rate. Second, Walk selection stabilises at approximately the same values, but has much higher variability and slower convergence. Also, interestingly, after an initial sharp increase the number of tip seen decreases again. This must be an anomaly caused by the fact that initially transactions are issued in 'generation cycles' which soon randomise into a proper population where each new transaction finds an increasing number of unapproved transactions until the situation stabilises.

This phenomenon indicates that the Walk policy accumulates a backlog in tips more slowly than URTS and could therefore be more robust in a setting with load fluctuations. We will need more and longer experiments to understand the dynamics in their full depth.

6 Conclusions

In this paper we have investigated the performance of two commonly used tip selection algorithms in IOTA, the uniform random tip selection (URTS) and the Walk tip selection. We have developed TangleSim, a simulator based on OMNeT++ which allows to compute metrics such as the time as a tip of a transaction, the time until approval and the number of tips seen by a transaction.

We find that the more complex walk tip selection method leads to faster approval of transactions and hence is less likely to accumulate a large backlog of transactions. However, on the long run, the number of tips seen seems to be similar for both policies. This is a very interesting result as the walk tip selection also is better with respect to the tangle stability, which was not discussed in depth in this paper, but can be derived from the transient analysis shown in Fig. 11.

There are still many open issues in this field. The question of scalability must be explored further and what the effects of changes in the load on the transaction throughput are. It seems as if high load is beneficial to the tangle consensus method, but it is unclear whether this observation holds beyond the considered range in load and time.

Acknowledgements We would like to thank the anonymous reviewers for their helpful comments, especially regarding comparison with the parameter settings in IOTA.

References

1. Nakamoto, S.: Bitcoin: a peer-to-peer electronic cash system. https://bitcoin.org/bitcoin.pdf (2008)
2. Buterin, V.: Ethereum: a next-generation smart contract and decentralized application platform. http://ethereum.org/ethereum.html (2017)
3. Duffield, E., Diaz, D.: Dash: A Privacy-Centric Cryptocurrency (2014)
4. statista.com.: PayPal's net number of payments from 1st quarter 2014 to 2nd quarter 2018. https://www.statista.com/statistics/218495/paypals-net-number-of-payments-per-quarter/. Accessed on Aug 2018
5. Popov, S.: The tangle. https://iota.org/IOTA_Whitepaper.pdf (2018)
6. Kusmierz, B., Staupe, P., Gal, A.: Extracting tangle properties in continuous time via large-scale simulations. https://www.iota.org/research/academic-papers (2018)
7. Gal, A.: The Tangle: An Illustrated Introduction, Part 3. https://blog.iota.org/the-tangle-an-illustrated-introduction-f359b8b2ec80 (2019)
8. Gervais, A., et al.: On the security and performance of proof of work blockchains. In: Proceedings of the 23nd ACM SIGSAC Conference on Computer and Communication Security (CCS). ACM (2016)
9. LeMahieu, C.: Nano: a feeless distributed cryptocurrency network. https://nano.org/en/whitepaper (2017)
10. Duffield, E., Diaz, D.: Dash: A Privacy-Centric Cryptocurrency (2014)
11. Popov, S., Saa, O., Finardi, P.: Equilibria in the tangle. In: arXiv e-prints, arXiv:1712.05385, Dec 2017
12. Varga, A., Hornig, R.: An overview of the OMNeT++ simulation environment. In: Proceedings of the 1st International Conference on Simulation Tools and Techniques for Communications, Networks and Systems & Workshops. Simutools '08. ICST, Marseille, France. ISBN: 978-963-9799-20-2 . http://dl.acm.org/citation.cfm?id=1416222.1416290 (2008)
13. Kusmierz, B.: The First Glance at the Simulation of the Tangle: Discrete Model (2017)

Committing to Quantum Resistance, Better: A Speed-and-Risk-Configurable Defence for Bitcoin Against a Fast Quantum Computing Attack

Dragos I. Ilie, William J. Knottenbelt and Iain D. Stewart

Abstract In light of the emerging threat of powerful quantum computers appearing in the near future, we investigate the potential attacks on Bitcoin available to a quantum-capable adversary. In particular, we illustrate how Shor's quantum algorithm can be used to forge ECDSA based signatures, allowing attackers to hijack transactions. We then propose a simple commit–delay–reveal protocol, which allows users to securely move their funds from non-quantum-resistant outputs to those adhering to a quantum-resistant digital signature scheme. In a previous paper (Stewart et al. R. Soc. Open Sci. 5(6), 180410 (2018)) [1] we presented a similar scheme with a long fixed delay. Here we improve on our previous work, by allowing each user to choose their preferred delay–long for a low risk of attack, or short if a higher risk is acceptable to that user. As before, our scheme requires modifications to the Bitcoin protocol, but once again these can be implemented as a soft fork.

1 Introduction

Bitcoin [2] is the first scalable and widely adopted decentralised cryptocurrency. Unlike most fiat currencies, decentralised digital money are not regulated by a central entity. Participants to the network cooperate by following a protocol to broadcast, validate, and record transactions in a decentralised, distributed, database-like structure called the blockchain [3]. The primary mechanism for establishing consensus and guaranteeing the immutability of transactions is called Proof of Work (PoW) and it fundamentally relies on hashes [4]. The other crucial cryptographic primitive

D. I. Ilie · W. J. Knottenbelt · I. D. Stewart (✉)
Centre for Cryptocurrency Research and Engineering, Imperial College London,
London SW7 2AZ, UK
e-mail: ids@imperial.ac.uk

D. I. Ilie
e-mail: dii14@imperial.ac.uk

W. J. Knottenbelt
e-mail: wjk@imperial.ac.uk

P. Pardalos et al. (eds.), *Mathematical Research for Blockchain Economy*,
Springer Proceedings in Business and Economics,
https://doi.org/10.1007/978-3-030-37110-4_9

is the digital signature scheme [5], which is used to selectively grant authorisation to change data (or move funds) on the blockchain. Currently, the public-key cryptography of choice in Bitcoin and most other blockchains is Elliptic Curve Cryptography (ECC) [6] whose security relies on the hardness of the Elliptic Curve Discrete Logarithm Problem (ECDLP), which is indeed classically intractable.[1] Therefore, Bitcoin and other similar blockchains provide decentralised and trustless peer-to-peer transactions with much stronger security, transparency, and ownership guarantees that banks or even governments can offer.

Quantum Computers (QCs) [7] appears as an idea in 1959 when Richard Feynman notices that the laws of quantum mechanics can be used to manipulate atoms or photons to perform highly efficient parallel computations [8]. However, the domain remains unexplored until further research continues in the 1980s and interest in the field slowly grows. One of the most exciting developments in quantum computing comes from Peter Shor in 1994 when he finds a quantum algorithm that threatens most of the popular public-key cryptography schemes (ECC included) [9]. As practical implementations of QCs remain a difficult engineering challenge due to quantum effects such as decoherence [10], it appears cryptographers will have enough time to upgrade their security to quantum resistant schemes.

The current implementations of QCs [11–13] do not have enough qubits to solve problems large enough to affect Bitcoin, but as more funding is being directed at this endeavour, different approaches for the architecture of QCs are being considered, tested, and implemented [14] so a sudden improvement might lead to a powerful QC appearing virtually overnight. Thus, Bitcoin and other blockchains will have to take action to allow their users to transition their funds to quantum resistant outputs even in the presence of a powerful QC.

In this paper, we aim to show the impact quantum algorithms have on Bitcoin and other blockchains relying on ECC, how it can be mitigated, and ultimately, how one can safely transition to quantum resistance. The remainder of the paper is structured as follows. Section 2 outlines the workings of Bitcoin and the relevant quantum computing attacks available. Section 3 examines the threats a slow or fast quantum-capable attacker poses to Bitcoin. Most importantly, in Sect. 4 we propose a protocol for transitioning from Bitcoin's current signature scheme to a quantum-resistant one chosen by the community. Finally, we present related work in Sect. 5 and conclude the paper in Sect. 6.

2 Background

In this section we present basic information on Bitcoin's signature scheme, on the quantum algorithms threatening Bitcoin, and on symmetric encryption which is required by our protocol. However, as each of these topics is by itself fairly complex, we recommend readers unfamiliar to these concepts to examine existing literature,

[1] The ECDLP is intractable only in specific groups, like the one used in Bitcoin.

such as: [2, 15, 16] for Bitcoin, [7, 17] for quantum computing, and [18] for quantum resistant schemes.

2.1 Bitcoin Fundamentals

Transactions: In Bitcoin, data is recorded in transactions built from inputs and outputs. Each input references a previously unreferenced output and provides proof of authorisation. The outputs that are unreferenced or "unspent" in a certain state of the blockchain constitute the Unspent Transaction Output (UTXO) set. Each output contains a challenge script which must be solved by any input wishing to reference it. Most commonly, a challenge script contains the hash of a public key pk and in order to spend it, an input must give this public key and a signature over the spending transaction that can be verified with pk. The act of providing this data serves as authorisation to spend the output and further secures the spending transaction as the signature can only be constructed with the secret key associated to pk, which will never be revealed to the network. In Bitcoin (and most other blockchains) the cryptography of choice is Elliptic Curve Cryptography (ECC), where the one-way (trapdoor) function between the secret key and the public key is exponentiation of elliptic curve points. Essentially, to deduce a secret key from a public key, one needs to solve the elliptic curve discrete logarithm problem (ECDLP) for which no efficient classical algorithm has been found. Therefore, the Elliptic Curve Digital Signature Algorithm (ECDSA) [19] gained popularity and was adopted by Bitcoin.

Blockchain and PoW: Once transactions are constructed by users, they are broadcast to other nodes in the network, until picked up by miners who validate them. Each miner, groups transactions in a block and competes with other miners to extend the blockchain, by solving a computationally intensive puzzle called Proof of Work. It is possible for multiple miners to extend on top of the same block at the same time, which would lead to a fork in the blockchain. Bitcoin imposes a rule specifying that only the longest chain is valid, which means that forks are automatically solved as one of the forked chains will be developed at a lower rate than the other, thus becoming invalid. Once in the blockchain, data is immutable as each block contains a hash of its predecessor, so any small change in a block breaks the link of hashes.

Hard/Soft Forks: Since Bitcoin first appeared in 2008, multiple protocol updates improved the security and user experience. In some scenarios the upgrade would invalidate existing rules so the blockchain irremediably splits into separate chains: one following the new protocol and one continuing to enforce the previous rules. This is called a hard fork as the two chains represent different cryptocurrencies at this point. Therefore, upgrading the core protocol is a sensitive issue because it forces miners to choose which chain they want to work on. Instead, protocol updates that are backwards compatible can be deployed through soft forks. If the set of transactions that are valid under the new rules is a subset of the set of transactions that were valid before the update, the protocol can be deployed as a soft fork. Miners and users who upgrade can use the new features, while old clients treat all transactions as valid and have the option to upgrade at a later time.

2.2 Quantum Computing and Algorithms

Quantum computing depends on several phenomena and laws of quantum mechanics [20] that are fundamentally different from those encountered in the day to day life. In general, quantum computations make use of superposition to encode many possible values on the same qubits[2] and then perform computations obtaining all the possible answers. However, when the system is measured, the superposition collapses with certain probabilities for each answer, losing all the other solutions. To overcome this, quantum algorithms use the underlying structure of the problem and manipulate the superimposed state in order to increase the likelihood of a certain result which can then be interpreted conclusively.

Shor's Algorithm can be used to solve the ECDLP and many other forms of the hidden subgroup problem [21], being the most urgent threat to most public-key cryptography in use. It works by preparing a superposition of states where each state is formed by concatenating x (the secret key) with the value of $f(x)$(the public key obtained by exponentiation). Then the Quantum Fourier Transform circuit [22] can be used to extract the period of the function f and subsequently compute x for any given $y = f(x)$. Shor's algorithm gives a polynomial-time attack against ECC [9] and most public key cryptography.

Grover's Algorithm tries to find a unique or very rare input value x to a black box function that produces a desired value $v = f(x)$ [23]. However, the time complexity is $O\left(\sqrt{\frac{N}{t}}\right)$ where N is the size of the domain of f and t is the number of solutions [24]. Hence, hashing rates do not improve considerably unless the range of solutions is very large.

Although such algorithms are impressive, quantum computers can only handle a few qubits at the moment and research into cryptographic schemes that appear to be quantum resistant is substantial (e.g. lattice-based, hash-based, code-based) [18]. Currently, the reason they are not popular is due to very large key sizes and signatures.

2.3 Symmetric Encryption

To implement our protocol, we make use of encryption which until now did not contribute to Bitcoin's security, so we recommend those who want to familiarise themselves with specific implementations and schemes to consult domain literature: [25, 26].

---

[2]Qubits are the equivalent of bits for QCs.

The basic idea behind encryption is to take a certain message (the plaintext) and pass it through a series of mathematical operations to obtain an output (the ciphertext) which does not reveal any information about the original message. Usually such algorithms work by taking a message and a key and returning a ciphertext which can only be decoded using the same key. This is called symmetric encryption and one example of its implementation is the Advanced Encryption Standard (AES) [25] selected by NIST. In fact, this algorithm is already used in Bitcoin to encrypt stored private keys. AES produces ciphertexts equal in length to the input plaintext and keys can be 128, 192, or 256 bits.

3 Quantum Computing Impact on Bitcoin

Having seen the main mechanisms that secure transactions and the quantum algorithms that are relevant in this context, we can theorise several attack vectors against Bitcoin that would be enabled by quantum computers.

3.1 Mining with Grover's Algorithm

Besides the urgent weakness of replacing ECDSA, we could consider possible attacks against hashing. Grover's algorithm does not offer considerable speedup for breaking the pre-image of hashed public keys found in challenge scripts. On the other hand, the Proof of Work puzzle that miners compete to solve is completely reliant on hashing power. The competition is to find a $nonce$ such that concatenated to a message m and hashed produces a number smaller than some target: $H(m||nonce) < target$ [2]. Classically, the brute force approach is the most efficient one, but using Grover's quantum algorithm we can accomplish this in $O\left(\sqrt{\frac{N}{target}}\right)$ time complexity, which means a more than quadratic speed-up that leads to an increased hash rate. However, current miners use optimised hardware (ASICs) [27] with machines working in parallel, so it is difficult to predict if or when QCs will be large and fast enough to outperform them. Therefore, in this paper, we will not address vulnerabilities rooted in Proof of Work as they do not seem to disrupt the network.

3.2 Solving the Elliptic Curve Discrete Logarithm Problem

When efficient QCs with large states will be physically realised, Bitcoin's Elliptic Curve Cryptography can be undermined by using Shor's quantum algorithm. In fact, an attacker with a quantum computer of about 1500 qubits [28, 29] can solve the ECDLP and compute an ECDSA secret key given the public key. Once the attacker deduces the secret key, he is indistinguishable from the original owner and he can successfully sign transactions consuming any UTXOs secured by that public key. Therefore, we outline the impact of this attack on Bitcoin, some protective measures users can take to secure their coins until spending, and how live transaction hijacking affects the transition to quantum resistance.

Public Key Unveiling: For the purpose of this analysis, it makes sense to distinguish between a slow QC and a fast one. Let us first, assume that a slow quantum computer is developed. While it can be used to solve the ECDLP, the computation lasts longer than the time it takes for a transaction to be included in the blockchain. Under this assumption, a QC is capable of deducing the private key from a formerly revealed public key. Hence, the first measure Bitcoin users can implement is to ensure all of their funds are secured by not yet revealed public keys. Bitcoin UTXOs with the `P2PK` type challenge script display the public key in the output of the transaction and although these type of challenge scripts are legacy, they currently secure about 1.77 million BTC [1]. Furthermore, instances of revealed public keys can arise from solving any type of challenge script. If an UTXO secured by *pk* is consumed and *pk* also secures other UTXOs which are not referenced in this transaction, they would become vulnerable as *pk* was revealed.[3] To prevent this attack users are advised to not reuse their public keys, which is in fact recommended behaviour in Bitcoin [30, 31]. However, about 3.9 million BTC [1] still reside in UTXOs that are secured by public keys revealed in some other input. Overall, these two cases of public key unveiling compromise at least 33% of all BTC, so we strongly advice users to move funds locked in such outputs immediately. Furthermore, publishing the key on a Bitcoin fork (e.g. Bitcoin Cash [32], Bitcoin Gold [33]) or as part of signed messages in forums, or in payment channels (e.g. Lightning Network [34]), would also reveal public keys, but estimates for the impact are harder to obtain.

3.3 Transaction Hijacking

Until now, we have assumed that only slow QCs are available, but the most powerful attack against Bitcoin can only be performed with a fast QC, that can deduce ECDSA secret keys faster than a new transaction can be inserted in the blockchain. Equipped with such technology, an attacker can successfully perform live transaction hijacking. Immediately after a transaction *TX* is broadcast, the attacker uses the now revealed public keys from the inputs of *TX* to compute the associated private keys. He, then creates a second transaction *TX'* that consumes the same outputs as *TX*, but sends the funds to an address under the attacker's control. Note that spending the same UTXOs is possible because the attacker is in control of all the private keys needed to generate valid signatures. If this procedure can be performed before *TX* is on the blockchain and *TX'* has a higher fee, miners will choose to include *TX'* and invalidate *TX* as they gain more. We call this attack live transaction hijacking as the original

[3]Consuming an UTXO secured by *pk* necessarily reveals *pk* in order to verify the signature.

transaction is overwritten by the malicious one. However, the success probability of such an attack is dependent on the performance of QCs, so perhaps the Bitcoin community will have enough time to protect against it.

3.4 Hindering Transition to Quantum Resistance

Assuming the above scenario becomes increasingly recognised by the Bitcoin community, a quantum resistant cryptographic scheme will be chosen to replace ECC. This upgrade can be deployed through a soft fork by repurposing an unused opcode[4] (`OP_CHECKQRSIG`) to create challenge scripts secured by quantum resistant public keys. The new feature, would allow users to move their ECDSA secured funds to UTXOs protected by the new cryptographic scheme introduced. However, once fast QCs are believed to have appeared, the community must invalidate all spending from ECDSA based challenge scripts as they are susceptible to live transaction hijacking. On the other hand, these funds are recoverable even in the presence of a fast QC if the appropriate transition protocol is deployed.

4 Protocol Specification

In this section, we present a scheme for transitioning Bitcoin to quantum resistance securely. We list our assumptions, the new consensus rules that need to be enforced, and examine each step of the process from both the user and the miner's perspective.

We note that quantum resistant digital signatures are required in order to deploy our scheme, so as explained in Sect. 3.4, they must have already been deployed in Bitcoin. Many people, perhaps a majority, might have already transitioned their funds via standard transactions while QCs were still nonthreatening and live transaction hijacking was not considered a risk. However, there will remain some ECDSA secured UTXOs for which the public keys are not revealed yet. For the sake of recovering these UTXOs, we must deploy a protocol update which invalidates ECDSA signatures as they currently are because they can be forged.

4.1 Upgraded Consensus Rules

We need some cryptographic tools that will contribute to the implementation of the protocol we propose:

[4]Opcodes are used in challenge scripts to perform any operations such as: hashing, signature verification, addition, etc. There are several unused opcodes reserved for extending the capabilities of challenge scripts.

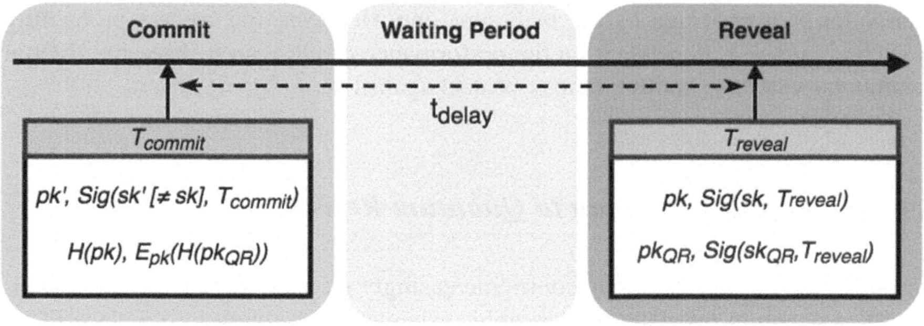

Fig. 1 Overview scheme of the protocol. Data in black can be added using existing functionality in Bitcoin, while the data in pink will be added in a segregated area. The waiting period, $\texttt{t}_{\texttt{delay}}$ is user configurable, but should be considered carefully as described in Sect. 4.4

1. H a cryptographic hash function.
2. E_k a symmetric encryption function with key k.

Using these constructs, we suggest the following upgrades to the consensus model. Rules for valid transactions remain unchanged, except for the verification of ECDSA signatures. A signature SIG over a transaction T_{reveal} against a public key pk, is valid only if:

1. *SIG* is valid according to the previous rules; i.e. *pk* can be used to verify that *SIG* really signs T_{reveal}.
2. T_{reveal} contains a quantum resistant signature (*QRSIG*) and public key (pk_{QR}) and pk_{QR} can verify that *QRSIG* signs exactly the same data as *SIG* does.
3. There exists a previous transaction T_{commit} which contains the following data: $(H(pk), E_{pk}(H(pk_{QR})))$, where $H(pk)$ is a tag and $E_{pk}(H(pk_{QR}))$ is the validation data.
4. T_{commit} is the first transaction with tag $H(pk)$ for which the validation data successfully decrypts to $H(pk_{QR})$ using pk as decryption key. T_{commit} is called a first valid commitment.

We describe in detail the structure of a first valid commitment and the reasoning behind its format in Sect. 4.3 (Fig. 1).

4.2 Overview

As explained in Sect. 3.4, ECDSA signatures cannot protect transactions any longer without some additional data. Therefore, the proposed protocol tightens the rules of valid transactions mandating that for every ECDSA signature $sig(sk, data)$, generated with a secret key sk, that signs some data, the transaction must contain a second quantum resistant signature over the same data: $sig(sk_{QR}, data)$. Furthermore, a proof for the common ownership of sk and sk_{QR} must be given to the network without revealing sk_{QR}. To this end, users publish a sort of hash commitment that we call "first valid commitment", that secretly, uniquely, and irreversibly links their unrevealed ECDSA public key to a quantum resistant one which will act as a secure surrogate. Assuming (pk, sk) is a classical ECDSA key-pair and (pk_{QR}, sk_{QR}) is a quantum resistant one, a valid commitment that links these two key-pairs will contain the hash of pk and a hash of pk_{QR} that was symmetrically encrypted using pk as the encryption key.

$$(H(pk), E_{pk}(H(pk_{QR})))$$

Our upgrade also requires valid commitments to be the first associated to their respective tag. Thus, only one quantum resistant public key will be considered as valid surrogate for a committed public key. The aforementioned steps and the security of our scheme will be detailed in the following sections, but we first present what this protocol entails for users and miners.

From a user's perspective, the transition is divided in two stages: Firstly, users have to secure their ECDSA public keys by secretly linking them to quantum resistant surrogates. Secondly, the extra quantum resistant signatures and public keys are added to all transactions that consume ECDSA secured UTXOs. More specifically, assume Alice wishes to use her ECDSA key-pair (pk, sk) to sign a transaction which spends some UTXOs under her control. Further, assume that she also controls a quantum resistant key-pair (pk_{QR}, sk_{QR}). Alice computes the hash of pk: $H(pk)$, which acts as a tag for identifying the commitment linking pk to its surrogate. Next, she uses the symmetric encryption algorithm to encrypt a hash of pk_{QR} using pk as encryption key, thus obtaining: $E_{pk}(H(pk_{QR}))$. Note that Alice is the only one who knows pk at this time, so she is the only who can successfully encrypt something that will later be decrypted using pk. She proceeds to insert both the tag and the encrypted data in an `OP_RETURN` type output, possibly adding some fixed extra bytes that signal to miners that this data is meant as proof for common ownership. As this represents Alice's commitment that the quantum resistant key-pair is in her control, we call this transaction: T_{commit}. Once it is included in the blockchain, Alice waits for a certain period until she is confident that history rewriting attacks cannot rewind the blockchain beyond T_{commit}. During this time, Alice should be very careful not to reveal pk as attackers might be able to overwrite her commitment. Moreover, Alice must ensure she does not lose sk_{QR} as it is the only key which can be accepted as surrogate for sk, so funds would be locked. Assuming she is careful and has waited appropriately, Alice can now spend any UTXOs secured by (pk, sk), by also including the extra quantum resistant signatures against pk_{QR}. Thus, for every ECDSA signature generated with sk over some data D, Alice also generates a quantum resistant signature using sk_{QR} over D and includes this signature and pk_{QR} in a segregated area of the transaction similar to how SegWit [35] was implemented. Notice how our protocol upgrade imposes no restrictions on the format of the transaction so the funds can be directed to any destination. Thus, Alice can send her funds to outputs protected by quantum resistant public keys under her control, successfully transitioning to quantum resistance.

From a miner's perspective, the first major change is rejecting any ECDSA signatures which do not have an associated quantum resistant surrogate or which are invalid anyway. Next, miners have to check the quantum resistant signature and the proof of common ownership, which entails finding the first valid commitment associated to the hash of the ECDSA public key pk that is required in the input. Miners compute $H(pk)$, and chronologically look for `OP_RETURN` type outputs which contain this tag. We recommend implementation of this search to be done via an index of surrogates similar to the one for the UTXO set. Miners would maintain a map from tags to ordered lists of various encrypted data, of which only the first to decrypt successfully matters as all the other are invalid attempts from attackers or spammers. Thus, miners iterate through the list of possible ciphertexts linked to the tag and try to decrypt each one. The first successful decryption is matched against the hash of the quantum resistant public key and if successful, the proof of common ownership is valid. When building the index, miners must be careful to parse the blockchain starting from the first block, otherwise they might miss valid commitments.

4.3 First Valid Commitment

In this section, we present some considerations around the creation, structure, processing, and security guarantees of a first valid commitment. We will use the term "unsolved commitment" to refer to commitments which have not been validated yet.

Creating a commitment is not always as trivial as inserting a transaction in the blockchain. After the protocol upgrade is deployed, transactions can either be created by spending quantum resistant UTXOs or by providing surrogates. Users with no such UTXOs and no surrogates already committed, will have to seek other means to insert their commitment in the blockchain. For instance, they could rely on users who have already transitioned to publish their tag and validation data.

The structure of a first valid commitment should allow miners to easily identify and index the surrogates. More importantly, if commitments are smaller than 80 bytes, they could fit in an `OP_RETURN` type output [36], requiring no modifications to the current Bitcoin code. This is a significant advantage as users can start committing even before the protocol update is deployed. To this end, we suggest the following format:

1. 2 byte flag chosen by the community to indicate the use of our protocol. Miners check this flag to distinguish between commitments and other uses of `OP_RETURN`.
2. 32 byte tag ($H(pk)$) used by miners as key for indexing the validation data. The community, needs to choose H such that the output is 256 bits long. For instance, a suitable cryptographic hash function that is already implemented in Bitcoin is SHA-256 [37].
3. 32 to 46 byte validation data ($E_{pk}(H(pk_{QR}))$). The remaining 46 bytes can be filled with the validation data. Encrypting pk_{QR} un-hashed would achieve the

same security, but would take more space because quantum resistant public keys are currently quite large. $H(pk_{QR})$ is only 32 bytes and its encrypted form might be a bit larger, but the community can choose an encryption algorithm with low or no ciphertext expansion [25].

Miners process a first valid commitment whenever the protocol specific flag appears in an OP_RETURN type output. The tag $H(pk)$ uniquely identifies the public key for which the commitment is intended, hence miners build an index of unsolved commitments: a map from tags to lists of validation data ordered chronologically. As commitments might have been published before the protocol is deployed, validators should parse the full blockchain. To check the proof of common ownership for pk and pk_{QR}, miners compute $H(pk)$, retrieve the ordered list of unsolved commitments, and try to decrypt each one using pk as decryption key. The first data successfuly decrypted is compared to the hash of pk_{QR} and if there is a match, the proof of common ownership is valid. Considering that any public key is irreversibly linked to only one quantum resistant surrogate, miners can replace the list indexed by $H(pk)$ with $H(pk_{QR})$ after the first time a proof is verified. With this approach, uses of pk in different transactions will require only the first proof of common ownership to iterate over validation data and decrypt possible commitments, while subsequent proofs can be verified with only one look-up in the index.

The security guarantees provided by the first valid commitment ensure that attackers cannot replace the intended quantum resistant public key with one under their control or hinder a user's transition to quantum resistance. To verify these claims, we examine the relevant actions available to an attacker at each stage of the transition.

Commitment Step—Only the user knows pk and he has just broadcast a transaction containing: (tag, validation_data). The tag is the output of a hash so breaking the pre-image resistance [4] to compute pk is impossible and the validation data is encrypted with pk which only the user knows. Hence, the only attacks made available by this act is stealing the tag or the validation data. The latter contains a quantum resistant public key out of the attacker's control, so it presents no interest. On the other hand, the tag could be used to create fake commitments of the form: (tag, random_data). Many malicious transactions could be inserted in the blockchain before the original one is added. However, none of them would be valid because the random data will fail to decrypt when miners check the proof of common ownership, so this does not affect the user in any way. Most of the damage would be sustained by the miners who have to index all the fake commitments and decrypt all the pieces of random data. These attacks are very expensive as each transaction has some associated fee, but in case they become popular, miners can increase the minimum fee required for a commitment to discourage attackers.

Waiting Period—Only the user knows pk and his transaction is included in the blockchain. Attackers have the same options as before. Furthermore, they can continue to publish fake commitments to flood the index with useless data forcing miners to increase fees and rendering the transition to quantum resistance a luxury on the short term. However, to sustain such an attack without any direct financial gains seems improbable.

Reveal Step—The user reveals pk as part of transitioning. Attackers are able to compute sk and forge ECDSA signatures, but the quantum resistant signature is still secure. Another action that was not available before is creating malicious valid commitments $(H(pk), E_{pk}(H(pk_{attacker})))$ using the revealed pk. These can be inserted in the blockchain at the current height, but miners will check the commitments in chronological order, hence the first valid one will be the original. Assuming the user has waited enough, attackers cannot rewind the blockchain to insert their valid commitment in front of the original. Therefore the proof of common ownership cannot be forged either.

Transition Complete—The reveal transaction is included in the blockchain. There might still be other UTXOs locked by pk, but given that commitments never expire, the user can transition the remaining UTXOs whenever he wishes because the link between pk and pk_{QR} is immutable. Under these conditions an attacker cannot replace the quantum resistant surrogate with his own, so the user's funds have safely transitioned. Note also that pk_{QR} can be used as surrogate for multiple ECDSA public keys, although such behaviour would inadvertently reveal potentially sensitive information about the user, i.e. the ECDSA public keys were controlled by the same owner.

4.4 Configurable Delay Considerations

The waiting period between committing a public key pk and showing it as part of solving a challenge script when transitioning funds is a crucial element of the scheme. Revealing pk, allows attackers to create their own valid commitments: $(H(pk), E_{pk}(H(pk_{attacker})))$. If the original commitment (T_{commit}) is very recent, attackers could insert their own in front of T_{commit} by rewinding the blockchain just enough blocks. Therefore, the user must reveal pk only when he believes quantum-capable attackers would not be able to rewind such a long history of the blockchain. In a previous paper [1] we justified the choice for a long waiting period and proposed a similar scheme which actually enforces a 6 months delay before spending is allowed. However, the current protocol offers more flexibility: users can reveal earlier if they have reasons to believe QCs will not overwrite their commitment. For example, low value transactions can have shorter waiting periods because an attack would not be profitable.

4.5 Quantum Resistant Surrogate

In this section, we discuss the format of the segregated area and explain how this change can be deployed as a soft fork. The purpose of the quantum resistant signature and public key is to replace their non quantum resistant analogues. For backwards compatibility, the ECDSA signature and public key are still required to satisfy clients who did not upgrade and follow the old consensus rules. The implementation we propose is open for debate, but we believe the quantum resistant data should be added in a segregated area of the transaction similarly to how SegWit is implemented [35]. The specific format of the data should be robust, unambiguous, and flexible enough to accommodate multi-sig [15] validation and other types of challenge scripts. We suggest to structure the surrogate data in two tables: one from public keys to their surrogates and one from ECDSA signatures to their quantum resistant replacements. Thus, for each public key pk that successfully verifies a signature, there must be an entry which is composed of pk, acting as a tag, and a quantum resistant public key. Similarly, for each signature sig that is verified, there must be an entry containing a tag (e.g. $H(sig)$, sig, or any unique identifier) and a quantum resistant signature that signs the same data. We sketch the new format of a transaction below:

- inputs
- outputs
- segregated area—for quantum resistant data
  - array—for mapping public keys to their quantum resistant surrogates
    pk—one of the public keys from the inputs
    pk_{QR}—the quantum resistant surrogate meant for pk
  - array—for mapping signatures to their surrogates
    $tag(sig)$—the unique identifier for sig
    sig_{QR}—the quantum resistant surrogate of sig.

We suggest this format because it stores each quantum resistant public key only once even if it is required for multiple signatures. As mentioned in Sect. 2, quantum resistant schemes have large keys and signatures, so being space efficient is important. Miners who upgrade to the new protocol, will validate transactions according to the rules introduced in Sect. 4.1. Using a design similar to SegWit, deployment of the protocol upgrade can be achieved as a soft fork. Old clients receive transactions stripped of the segregated area and verify only the ECDSA signatures, while upgraded clients fully validate the segregated area as well.

5 Related Work

In this section, we would like to give credit to some other ideas which address the same problem our paper focuses on, i.e. transitioning Bitcoin to a post-quantum world in the presence of an already-fast quantum-capable attacker.

The first mention we could find of a scheme transitioning Bitcoin to quantum resistance is made by Adam Back, referring to an informal proposal by Johnson Lau [38]. The idea presented is to publish a hash commitment, $H(pk, pk_{QR})$, and to later reveal (pk, pk_{QR}) and provide quantum resistant signatures against pk_{QR} that sign the same data as the ECDSA signatures against pk. However, we could not find further details on this scheme and it seems there might be some security issues around the delay period between the two phases. More specifically, it seems possible for an attacker to wait for a reveal transaction, insert his own commitment: $H(pk, pk'_{QR})$, and then reveal (pk, pk'_{QR}), all of this before the original reveal transaction has been confirmed. It would now be impossible for any miner, to decide which reveal transaction is invalid as they both have valid commitments.

Another scheme is more formally described by Tim Ruffing in the Bitcoin-dev mailing list [39]. It leverages the idea of committed transactions to only allow ECDSA transactions that have already been committed to in the form of encrypted data. To protect against attacks around the delay period, this scheme enforces a first valid commitment rule. However, the tag used to identify commitments is the challenge script, not a certain public key, so there could be cases where the scheme reveals public keys which also secure other challenge scripts which have not been committed yet, thus allowing attackers to hijack those outputs. Furthermore, this scheme relies on being able to create a transaction without revealing the public keys needed to any parties. However, multi-sig type outputs require multiple signatures and public keys from different parties, so public keys will need to be revealed in order to create the spending transaction. Our scheme overcomes this issue by operating directly on the abstraction level of individual signatures, rather than full transactions.

Finally, we mention another design very similar to Johnson Lau's, but formally described with more attention to attacks around the delay period. We have previously coauthored a paper [1] describing this proposal. Although effective, the main drawback of this scheme is the considerable delay that we had to enforce in order to ensure no history rewinding attacks are possible. In fact, the motivation for our current work is exactly the feedback we received on this large delay period, therefore, our new proposal grants users the right to choose their own delay period (and associated level of chain-rewind risk).

6 Conclusion

Taking into account the increasingly probable scenario of a powerful quantum computer being physically implemented in the near future, we have outlined how Bitcoin is susceptible to live transaction hijacking as a consequence of exposed elliptic curve public keys. To mitigate against such attacks we have proposed a commit–delay–reveal scheme that enables the secure transition of funds to quantum-resistant outputs. The protocol allows users to execute the first step of transitioning even before the upgrade is deployed as the necessary functionality already exists in Bitcoin. Code changes are required only for the reveal stage of the transition, and they can be implemented as a soft fork, allowing users to upgrade at their own convenience. Furthermore, the format of the reveal transaction is not restricted in any form by the proposed changes, thus users can spend and create any types of challenge scripts. As an improvement to our previous work [1], the current protocol allows each user to choose his preferred delay period between committing and revealing. However, we emphasise the trade-off between the speed and risk of transitioning. A faster transition will result in less blocks that need to be rewound by an attacker, therefore the risk of a chain-rewind attack increases. Ultimately, users should decide on the duration of the waiting period once the capabilities of future quantum computers are better understood and performance estimates become more reliable.

References

1. Stewart, I., Ilie, D., Zamyatin, A., Werner, S., Torshizi, M., Knottenbelt, W.J.: Committing to quantum resistance: a slow defence for bitcoin against a fast quantum computing attack. R. Soc. Open Sci. **5**(6), 180410 (2018)
2. Nakamoto, S.: Bitcoin: a peer-to-peer electronic cash system. https://bitcoin.org/bitcoin.pdf (2008). Accessed 01 July 2015
3. Crosby, M., Pattanayak, P., Verma, S., Kalyanaraman, V., et al.: Blockchain technology: beyond bitcoin. Appl. Innov. **2**(6–10), 71 (2016)
4. Rogaway, P., Shrimpton, T.: Cryptographic hash-function basics: definitions, implications, and separations for preimage resistance, second-preimage resistance, and collision resistance. In: International Workshop on Fast Software Encryption, pp. 371–388. Springer, Berlin (2004)
5. Rivest, R.L., Shamir, A., Adleman, L.: On digital signatures and public-key cryptosystems. Technical Report, Massachusetts Inst of Tech Cambridge Lab for Computer Science (1977)
6. Miller, V.S.: Use of elliptic curves in cryptography. In: Williams, H.C. (ed.) Proceedings of the Advances in Cryptology—CRYPTO '85, pp. 417–426. Springer, Berlin, Heidelberg (1986)
7. Kaye, P., Laflamme, R., Mosca, M.: An Introduction to Quantum Computing. Oxford University Press (2007)
8. Feynman, R.: Theres plenty of room at the bottom. In: Feynman and computation, pp. 63–76. CRC Press (2018)
9. Shor, P.W.: Polynomial-time algorithms for prime factorization and discrete logarithms on a quantum computer. SIAM Rev. **41**(2), 303–332 (1999)
10. Schlosshauer, M.A.: Decoherence: And the Quantum-to-Classical Transition. Springer Science & Business Media, Berlin (2007)
11. Debnath, S., Linke, N.M., Figgatt, C., Landsman, K.A., Wright, K., Monroe, C.: Demonstration of a small programmable quantum computer with atomic qubits. Nature **536**(7614), 63 (2016)
12. Veldhorst, M., Yang, C., Hwang, J., Huang, W., Dehollain, J., Muhonen, J., Simmons, S., Laucht, A., Hudson, F., Itoh, K., et al.: A two-qubit logic gate in silicon. Nature **526**(7573), 410 (2015)
13. Watson, T., Philips, S., Kawakami, E., Ward, D., Scarlino, P., Veldhorst, M., Savage, D., Lagally, M., Friesen, M., Coppersmith, S., et al.: A programmable two-qubit quantum processor in silicon. Nature **555**, 633–637 (2018)
14. Bettelli, S., Calarco, T., Serafini, L.: Toward an architecture for quantum programming. Eur. Phys. J. D-At. Mol. Opt. Plasma Phys. **25**(2), 181–200 (2003)
15. Antonopoulos, A.M.: Mastering Bitcoin: Unlocking Digital Cryptocurrencies. O'Reilly Media, Inc. (2014)
16. Narayanan, A., Bonneau, J., Felten, E., Miller, A., Goldfeder, S.: Bitcoin and cryptocurrency technologies. Princeton University Press (2016)

17. Nielsen, M.A., Chuang, I.: Quantum computation and quantum information. Cambridge University Press (2000)
18. Bernstein, D.J., Lange, T.: Post-quantum cryptography. Nature **549**(7671), 9 (2017)
19. Bitcoin community. Elliptic Curve Digital Signature Algorithm. https://en.bitcoin.it/wiki/Elliptic_Curve_Digital_Signature_Algorithm. Accessed 18 Feb 2018
20. Jogenfors, J.: Quantum bitcoin: an anonymous and distributed currency secured by the no-cloning theorem of quantum mechanics. arXiv:1604.01383 (2016)
21. Mosca, M., Ekert, A.: The hidden subgroup problem and eigenvalue estimation on a quantum computer. In: NASA International Conference on Quantum Computing and Quantum Communications, pp. 174–188. Springer, Berlin (1998)
22. Lavor, C., Manssur, L., Portugal, R.: Shor's algorithm for factoring large integers. arXiv: quant-ph/0303175 (2003)
23. Grover, L.K.: A fast quantum mechanical algorithm for database search. In: Proceedings of the twenty-eighth annual ACM symposium on Theory of computing, pp. 212–219. ACM (1996)
24. Boyer, M., Brassard, G., Høyer, P., Tapp, A.: Tight bounds on quantum searching. Fortschritte der Physik **46**(4–5), 493–505 (1998)
25. Daemen, J., Rijmen, V.: The design of Rijndael: AES-the advanced encryption standard. Springer Science & Business Media, Berlin (2013)
26. Elminaam, D.S.A., Abdual-Kader, H.M., Hadhoud, M.M.: Evaluating the performance of symmetric encryption algorithms. IJ Netw. Secur. **10**(3), 216–222 (2010)
27. Taylor, M.B.: The evolution of bitcoin hardware. Computer **50**(9), 58–66 (2017)
28. Proos, J., Zalka, C.: Shor's discrete logarithm quantum algorithm for elliptic curves. arXiv: quant-ph/0301141 (2003)
29. Tessler, L., Byrnes, T.: Bitcoin and quantum computing. arXiv:1711.04235 (2017)
30. Meiklejohn, S., Pomarole, M., Jordan, G., Levchenko, K., McCoy, D., Voelker, G.M., Savage, S.: A fistful of bitcoins: characterizing payments among men with no names. In: Proceedings 2013 Internet Measurement Conference, pp. 127–140. ACM (2013)
31. Schneider, N.: Recovering bitcoin private keys using weak signatures from the blockchain. http://www.nilsschneider.net/2013/01/28/recovering-bitcoin-private-keys.html (2013). Accessed 18 Feb 2018
32. Bitcoin Cash. https://www.bitcoincash.org/. Accessed 18 Feb 2018
33. Bitcoin Gold. https://bitcoingold.org/. Accessed 18 Feb 2018
34. Poon, J., Dryja, T.: The bitcoin lightning network. https://lightning.network/lightning-network-paper.pdf (2016). Accessed 07 July 2016
35. Lombrozo, E., Lau, J., Wuille, P.: BIP141: segregated witness (consensus layer). https://github.com/bitcoin/bips/blob/master/bip-0141.mediawiki (2012). Accessed 18 Feb 2018
36. Bitcoin community. OP_RETURN. https://en.bitcoin.it/wiki/OP_RETURN. Accessed 18 Feb 2018
37. Eastlake III, D., Hansen, T.: US secure hash algorithms (SHA and HMAC-SHA) (2006)
38. Adam Back. https://twitter.com/adam3us/status/947900422697742337. Accessed 18 Feb 2018
39. Ruffing, T.: https://lists.linuxfoundation.org/pipermail/bitcoin-dev/2018-February/015758.html. Accessed 18 Feb 2018

Neural Networks for Cryptocurrency Evaluation and Price Fluctuation Forecasting

Emmanouil Christoforou, Ioannis Z. Emiris and Apostolos Florakis

Abstract Today, there is a growing number of digital assets, often built on questionable technical foundations. We design and implement neural networks in order to explore different aspects of a cryptocurrency affecting its performance, its stability as well as its daily price fluctuation. One characteristic feature of our approach is that we aim at a holistic view that would integrate all available information: First, financial information, including market capitalization and historical daily prices. Second, features related to the underlying blockchain from blockchain explorers like network activity: blockchains handle the supply and demand of a cryptocurrency. Lastly, we integrate software development metrics based on GitHub activity by the supporting team. We set two goals: First, to classify a given cryptocurrency by its performance, where stability and price increase are the positive features. Second, to forecast daily price tendency through regression; this is of course a well-studied problem. A related third goal is to determine the most relevant features for such analysis. We compare various neural networks using most of the widely traded digital currencies (e.g. Bitcoin, Ethereum and Litecoin) in both classification and regression settings. Simple Feedforward neural networks are considered, as well as Recurrent neural networks (RNN) along with their improvements, namely Long Short-Term Memory and Gated Recurrent Units. The results of our comparative analysis indicate that RNNs provide the most promising results.

Keywords Cryptocurrency · Deep learning · Neural network · Blockchain · Price variation prediction · Coin features · Feature importance

E. Christoforou (✉) · I. Z. Emiris · A. Florakis
Department of Informatics and Telecommunications, National and Kapodistrian University of Athens, Athens, Greece
e-mail: echristo@di.uoa.gr

I. Z. Emiris
e-mail: emiris@di.uoa.gr

A. Florakis
e-mail: sdi1400217@di.uoa.gr

I. Z. Emiris
ATHENA Research and Innovation Center, Maroussi, Greece

P. Pardalos et al. (eds.), *Mathematical Research for Blockchain Economy*,
Springer Proceedings in Business and Economics,
https://doi.org/10.1007/978-3-030-37110-4_10

1 Introduction

Digital cipher currencies based on blockchains as introduced by Nakamoto's specification in 2008 [16] are exponentially growing in number, during the last few years [8], thus affecting significantly the global financial and trading scene. Today, there is a large number of different cryptocurrencies (more than 1,600). Blockchain is primarily responsible for most of the advantages of cryptocurrencies over fiat currencies, such as decentralization and anonymity. The increasing interest around blockchain-based currencies underlines the importance of methods to evaluate them and forecast their price tendencies. Our paper focuses on artificial neural networks, but first surveys related previous work.

There are several existing works for stock price forecasting with time series prediction methods, see e.g. [2], and for stock-market crisis forecasting using either deep and statistical machine learning, e.g. [5], or computing and manipulating copulas [4]. Moreover, neural networks have already been used to forecast stock and cryptocurrency price fluctuations in several works: in [15] they examined the accuracy of neural networks for the prediction of the direction of Bitcoin price (in USD) using a Bayesian optimized Recurrent neural network and a Long Short-Term Memory (LSTM) network, against the Autoregressive integrated moving average (ARIMA) model, with the LSTM having the highest classification accuracy (52% accuracy and 8% root mean squared error). Neural networks have been designed to examine whether chaoticity is inherently improving short-term predictability of cryptocurrencies [14]. Several technical indicators were included in deep learning architectures in order to predict the future return trend of Bitcoin [17], and in hybrid neural networks with generalized auto-regressive conditional heteroskedasticity (ANN-GARCH) to forecast Bitcoin's price volatility [13]. Blockchain features are explored as input to neural networks that may explain Bitcoin's price hikes using several machine learning methods [18], and as input to a Bayesian neural network to model and predict the Bitcoin price [10]. Other works use neural networks to predict the fluctuation of Bitcoin's price and transactions based on user opinions and sentiments derived by online forums [12], or experiment with Convolutional neural networks, trained with Deep Reinforcement Learning, to extract portfolio weights using historic prices of sets, with financial assets as input [11].

The main contribution of this work is to explore cryptocurrencies with neural networks in order to rank them as positive or negative, based on stability or price increase, as well as to capture their daily price tendency through regression. A related goal is to determine the most relevant features for such analysis. Blockchain-based currencies, are more accurately investigated by considering all of their relevant aspects namely financial, including market capitalization and historical daily prices, blockchain-related features, such as network activity, as well as software development metrics based on GitHub activity. Various neural networks are designed, trained, validated and experimentally compared looking at their accuracy and using most of the widely traded digital currencies. Simple Feedforward neural networks are considered, as well as Recurrent neural networks (RNN) along with their improvements, namely Long short-term memory (LSTM) and Gated recurrent units (GRU). The results of our comparative analysis are briefly reported in Tables 3 and 4 and indicate that RNNs provide the most promising results.

The rest of the paper is organized as follows. Section 2 outlines our data, their processing and the neural networks that are compared. Section 3 presents the results of the neural networks in classifying cryptocurrencies and forecasting price fluctuations. Section 4 offers a discussion of results, and future work.

2 Methods

In this section, we outline the data and their characteristics. We then outline the high-level theme of our analysis, the generation of the labels used for training, and the neural networks used.

2.1 Cryptocurrency Features

To describe efficiently daily price variations we consider several aspects.

Most of the existing approaches process time-series of prices and technical indicators; similarly this work also relies on historical daily prices. These prices consist of the open, high, low, and close (OHLC) prices, volume and market capitalization (Market Cap), all at a daily level.

Features representing technological aspects of a blockchain, at a daily level, are included to reflect price variations, considering that the blockchain records all transactions of a cryptocurrency. These data are collected from blockexplorers: platforms that allow search and navigation through the blocks of a blockchain, in order to produce several statistics. Some of these blockchain features reflect the daily number of blocks that were on the chain (block count), the number of bytes broadcast in final blocks (block size) and the difficulty level of the hash function to find a new block (average difficulty). Other features track the volume (in USD) that circulates on the blockchain per day (on-chain transaction volume), the number of transactions on the blockchain (transaction count), the USD value of the volume at cryptocurrency exchanges (exchange volume), and the number of new coins generated (generated coins). Finally, there are also features for the number of unique addresses used on the blockchain per day (active addresses), the total amount of fees (fees) and their median value (median fee), and the realized capitalization (realized cap), a metric that attempts to improve the market cap by counting the price when each coin lastly moved through the blockchain, instead of counting all of the mined coins at the last market price.

Table 1 Features collected for cryptocurrencies (financial features are in USD)

Scope	Features
Financial	Open, high, low, close, volume, market capitalization
Blockchain explorers	#active addresses, adjusted transaction volume, average difficulty, block count, block size, exchange volume, fees, #generated coins, median fee, median transaction value, payment count, realized cap, transaction count, on-chain transaction volume
Developer (Git)	#closed issues, #total issues, commit count (4 weeks frequency), #forks, #pull request contributors, #pull requests merged, #stars, #subscribers

In addition, other features describing the development activity, such as the commit count in a four-week interval, and the popularity of a cryptocurrency (number of stars, forks, subscribers and contributors) are also considered. These features are collected from the git repositories (most importantly, GitHub) of each cryptocurrency.

These 28 features (Table 1) are selected in order to identify those that affect or describe the price variations and they are collected from several websites such as CoinMarketCap (OHLCV prices), Coin Metrics (blockexplorer features) and CoinGecko (Git features).[1] All features are normalised feature-wise so that all inputs lie in [0, 1].

For classifying the cryptocurrencies one needs to generate appropriate labels: Each daily feature vector is labeled as "positive" or "negative", based on the variation of the closing price (p_t) in comparison to the mean price of the previous 30 days. When p_t is at least as large as 99% μ_{p_t}, namely

$$p_t \geq \mu_{p_t} - 1\% \, \mu_{p_t}, \quad \mu_{p_t} = \frac{1}{30} \sum_{i=1}^{30} p_{t-i},$$

we label the cryptocurrency as positive. Otherwise, it is negative, resulting to a dataset with 1035 positive and 760 negative instances for Bitcoin (BTC) in 2014–2018 (Fig. 1). We expect the neural network to identify those features that affect the price fluctuations, covering most aspects of a cryptocurrency (financial, blockchain, development). For the regression task, we aim to forecast price fluctuations, and set as target the closing price of the following day.

[1] https://coinmetrics.io, https://www.coingecko.com/en, https://coinmarketcap.com.

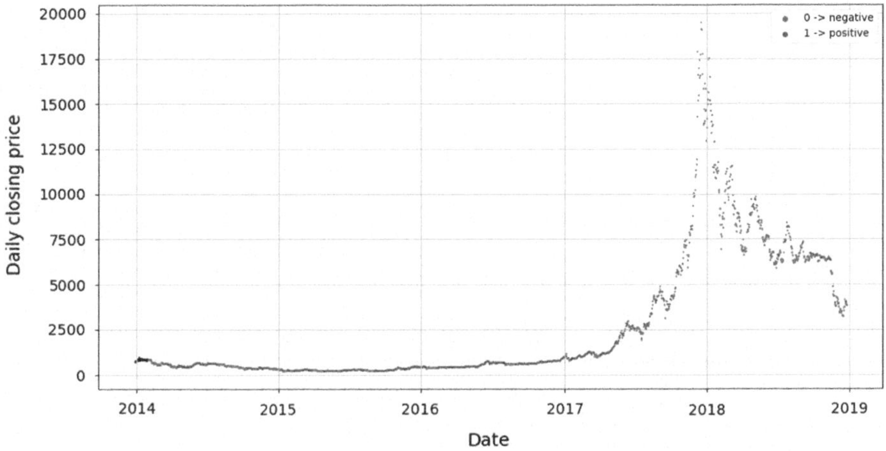

Fig. 1 BTC daily closing prices in 2014–18 labeled positive (green) or negative (red) (color figure online)

2.2 *Neural Networks*

Neural networks (NNs) are composed of an input layer, followed by an arbitrary number of hidden layers, and an output layer that makes the final decision or prediction. Trained on a set of input-output pairs, they model the correlation (or dependencies) between those inputs and outputs. A neural network is employed in two phases: The forward pass where the input signal flows from input through hidden layers, to the output layer. The latter is evaluated by the ground truth labels or values. Then, a backward pass follows, where the network parameters are back-propagated: the network weights and biases are adjusted in order to minimize error, using gradient-based optimization. The two passes are repeated until the loss does not reduce (convergence). Neural networks are distinguished in Feedforward neural networks (FNNs) and RNNs, based on whether they allow cyclic connections between nodes or not. Using cyclic connections, RNNs use internal state (memory) to process sequential data, such as time series.

Data points indexed by time may be processed as time series. FNNs are able to handle data only in a unified way, thus ignoring any underlying time-dependencies between their time steps. As a result, it becomes difficult for the network to identify hidden patterns that are related to time-dependencies. On the other hand, RNNs learn conditional dependencies between sequence elements during learning. RNNs process separately each step of the time series keeping a memory mechanism about the whole processing of the series. The content of this memory unit is used while processing each timestep. Since our data in this work are daily features for each coin, they can be processed both as independent instances in FNNs, as well as time series in RNNs.

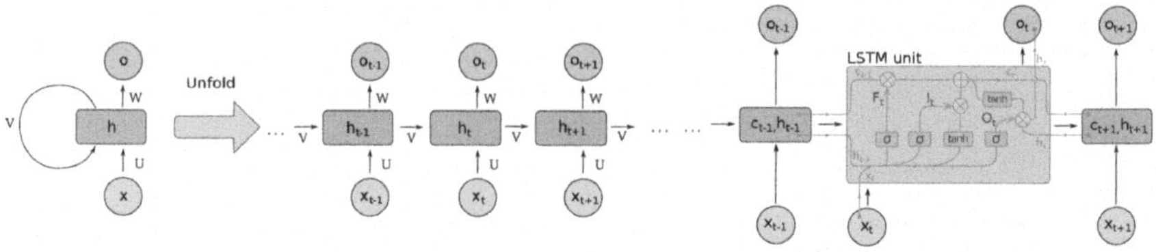

Fig. 2 Unfolding of a basic RNN (https://commons.wikimedia.org/wiki/File:Recurrent_neural_network_unfold.svg) and LSTM (https://commons.wikimedia.org/wiki/File:Long_Short-Term_Memory.svg)

Feedforward neural networks (FNNs). Deep FNNs, also known as multilayer perceptrons (MLPs), are neural networks with undirectional information flow, meaning that there are no cycles or feedback connections to feed back the output into the network. The basic idea behind an FNN is the perceptron (neuron), a linear classifier, that separates input into two categories by a straight line. An MLP is composed of more than one perceptrons placed in a single-direction (forward) graph. In this work we tested several such networks with an input layer, one to three hidden layers, and an output layer.

Recurrent neural networks (RNNs). For sequential data, where input is interdependent, FNNs appear to be inefficient. RNNs allow cyclic connections (Fig. 2), fitting better to dynamic processes, such as time series. These models are able to represent the relation between previous input-output pairs, since every new output is a function of the previous one. RNNs process one example at a time, retaining memory with contextual information, to be reused at the next time step. This recurrent formulation allows the RNN to share the same weights across several time steps. In theory, classic ("vanilla") RNNs can keep track of arbitrarily long-term dependencies in the input, but suffer from computational issues. During training, the back-propagated gradients may "vanish" (tend to zero) or "explode" (tend to infinity), because of the computations using finite-precision arithmetic. Therefore, a very deep computational graph of an RNN is unlikely to understand long-term dependencies. In order to cope with long-term dependency difficulties, Long Short-Term Memory (LSTM) and Gated Recurrent Units (GRU) architectures have been proposed. The idea behind LSTM and GRU units is to create connections through time with a constant error flow, thus the gradient neither explodes nor vanishes.

Long Short-Term Memory (LSTM). LSTM [9] is an RNN architecture. They were developed to deal with the vanishing gradient problems of traditional RNNs, providing a robust extension. LSTMs are explicitly designed to avoid long-term dependency problems. Their default behavior is to remember information for long time periods. This is why they are one of the most popular NNs for sequence learning, allowing gradients to flow unchanged, while preserving previous information. LSTM not only keep adjacent temporal information on a spontaneous manner, but also control long-term information introducing the notion of "controlling gate". A common LSTM unit is composed of a cell, an input gate, an output gate and a forget gate. The cell remembers values over arbitrary time intervals and the 3 gates regulate the flow of information into and out of the cell. Initially, the forget-gate extracts the amount of information that should be preserved from the prior state. Then, the input gate determines how much "current" information should be used as input in order to generate the current state. The output gate filters the information deemed significant. This procedure is repeated in every time step of sequential processing, allowing LSTM memory to remember or forget cell states.

Gated Recurrent Units (GRU). GRUs [6] are gating mechanisms in RNNs, similar to LSTMs with forget-gate, but have fewer parameters, as they lack an output gate. GRUs have been shown to exhibit better performance on certain smaller datasets [7]. They can be trained to remember information from long ago, without washing it through time, and to remove information which is irrelevant to the prediction by deciding what should be passed to the output. A GRU is composed of a cell, an update gate and a reset gate. The update-gate determines the previous time steps that need to be passed along to the future, while the reset-gate decides the past information to forget.

In our work we tested both of the above cell types. Sequences are fed as input to the RNNs in order to identify the underlying patterns. The predicted price or class is regarded as the output of our network.

3 Results

In this section, we outline the metrics used to evaluate our results both for classification and regression, while we outline the employed architectures. A summary of results is in Tables 3 and 4 with training and testing times, and the coins that were used. Our NNs were developed using the open-source software library TensorFlow [1]. Training was accomplished within minutes using the GPU accelerated environment of Google's Colaboratory.

3.1 Classification

True positives (TP) and true negatives (TN) are the outcomes where the model predicted the correct class (positive or negative correspondingly). Incorrect outcomes are defined as false positives (FP) and false negatives (FN). To evaluate the results of the classifiers we use the following metrics:

$$\text{Acc} = \frac{\text{TP}+\text{TN}}{\text{TP}+\text{TN}+\text{FP}+\text{FN}}, \quad \text{Prec} = \frac{\text{TP}}{\text{TP}+\text{FP}}, \quad \text{Rec} = \frac{\text{TP}}{\text{TP}+\text{FN}},$$

where Accuracy (Acc) indicates the total proportion of correct predictions among the total number of cases examined; Precision (Prec) is the fraction of true relevant instances predicted in a class; Recall (Rec) is the fraction of true relevant instances predicted over the total amount of relevant instances.

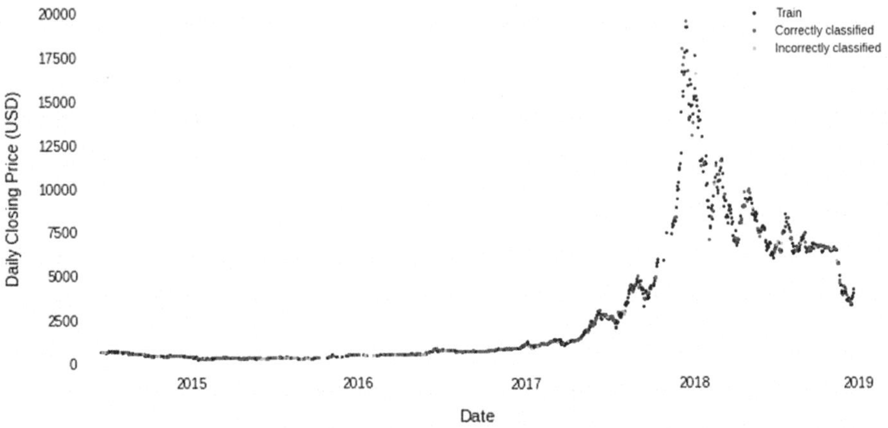

Fig. 3 Correctly (blue) and incorrectly (yellow) classified instances against closing price (BTC, 14/06/2014–21/12/2018), by an FNN with 3 hidden layers of 50 nodes (Acc: 84%, Prec: 86%, Rec: 89%). Black dots are training instances (color figure online)

FNN. An FNN is trained with historical instances of our dataset for Bitcoin from 14/06/2014 to 24/08/2017 and is tested using more recent instances (25/08/2017–21/12/2018). Using these instances as our test set, we want to investigate whether a plain NN is able to correctly classify them, identifying the underlying patterns that characterize the price trend of a cryptocurrency. A NN with three hidden layers of 50 nodes has Acc: 64% (while Prec: 56%, Rec: 76%).

Selecting the test set randomly from the whole sample and training an FNN with three hidden layers of 50 nodes, results to Acc: 84% (Prec: 86%, Rec: 89%), which is expected since the neural network is trained with samples through the whole history of the coin. Therefore, it is trained to understand most of its spectrum, classifying most of the instances correctly (Fig. 3).

Another approach is to divide every day's features by the corresponding value of the previous day, so as to represent the daily relevant change. Thus, we allow the NN to learn directly the rate at which features change, instead of their absolute values or differences between them. An FNN with one hidden layer of 100 nodes was trained with historical values for BTC (15/06/2014–09/12/2017) and tested using values on 10/12/2017–21/12/2018. It achieved Acc: 70%, Prec: 57%, Rec: 74% (Fig. 4). In Fig. 5 we observe the resulting labels on the test set, which are close to the original (Fig. 1).

RNN. Using the same features, we train a single layer RNN with a GRU cell of 100 nodes, where the input is time series of a 4-day time window (for each instance the previous 4 days are used). The first 70% of the dataset is used for training and the rest 30% for testing, achieving Acc: 71%, Prec: 67%, Rec: 71%. Also, a single layer RNN with an LSTM cell of 128 nodes and 10-days time window has been trained with data from BTC, Ethereum (ETH) and Litecoin (LTC) and was tested with a different cryptocurrency, namely Dash (DASH) coin instances. The results were Acc:

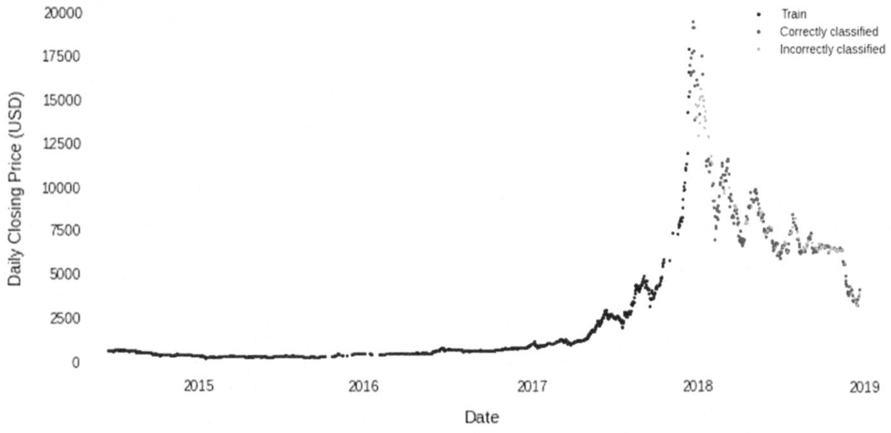

Fig. 4 Correctly (blue) and incorrectly (yellow) classified instances against closing price (BTC, 10/12/2017–21/12/2018), by an FNN with one hidden layer of 100 nodes (Acc: 70%, Prec: 57%, Rec: 74%). Black are training instances (color figure online)

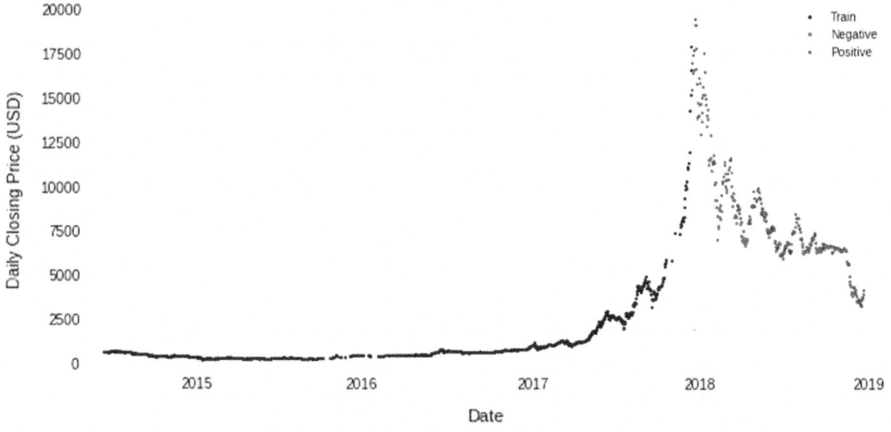

Fig. 5 Resulting labels (BTC, 10/12/2017–21/12/2018) by an FNN with one hidden layer of 100 nodes (Acc: 70%, Prec: 57%, Rec: 74%)

64%, Prec: 49%, Rec: 82%. It is very encouraging to notice that the performance is comparable to models tested on the same coin as the one used at training.

The previous RNN approaches exploit only past information for every instance. In order to make use of past and future states for each instance in a 4-day time window, we apply Bidirectional RNN (BRNN) [19]. The BRNN here consists of two stacked GRUs of 80 nodes with opposite directions (positive and negative time direction) to the same output. This enables the use of information from past and future states simultaneously. Hence, the performance may be enhanced by the prices tendencies located before and after every price in the specified window. In Fig. 6 the correctly and incorrectly classified instances are plotted, with Acc: 72%, Prec: 70%, Rec: 65%.

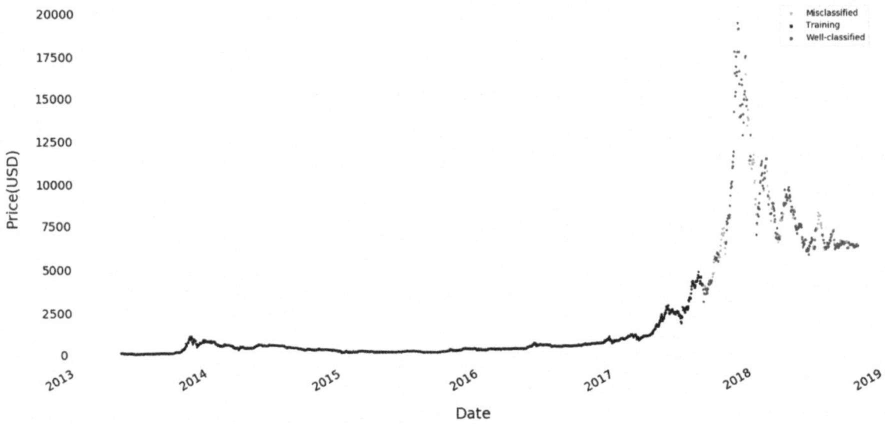

Fig. 6 Correctly (blue) and incorrectly (yellow) classified instances against the closing price of BTC, by BRNN with two GRU cell of 80 nodes (Acc: 72%, Prec: 70%, Rec: 65%). Black are training instances (color figure online)

FNN and RNN with single coin. Combining some of the previously mentioned layers we are able to build more effective neural networks. Here, we design a network with four hidden layers: a fully-connected layer of 64 nodes, a BRNN with LSTM layers of 128 nodes, followed by another LSTM layer of 128 nodes and a fully-connected layer with 64 nodes. As input we use the daily relevant change of each feature, as described in previous experiment. The network was trained using 10-day time-step window for BTC instances (15/06/2014 until 06/04/2018). The results on the test set (BTC 07/04/2018–06/04/2019) indicate that this network outperforms the previous experiments with Acc: 78% (Prec: 72%, Rec: 90%) (Fig. 7). Therefore, we use this model to explore the importance of each feature.

To identify the most important features in this work we use techniques such as permutation importance. Specifically, after training, we distort one of the features and we evaluate the accuracy of the model on the resulting test set. If the accuracy diverges significantly from our baseline, which is the accuracy of the same model on the original test set (Acc: 78%, Fig. 7), we conclude the corresponding feature is important. Repeating the same procedure for every feature allows us to distinguish those that appear to be more important.

We set two distortion test cases: (i) replace all values of a feature with a single value, indicating the feature does not change through time (called "Repeating-value importance"), and (ii) randomly shuffle all values of a feature ("Permutation importance"). The results are in Table 5. We report the accuracy of the model per feature, and the MAPE and MSE from the baseline accuracy, sorted by the MSE of Permutation importance. For Permutation importance, every experiment was performed 5 times, for accuracy, and the mean values are reported. MSE is used in order to take into account any outliers, especially for Permutation importance. Higher MAPE and MSE indicate the most important features. As we move to the top we find mostly financial and blockchain features, as expected.

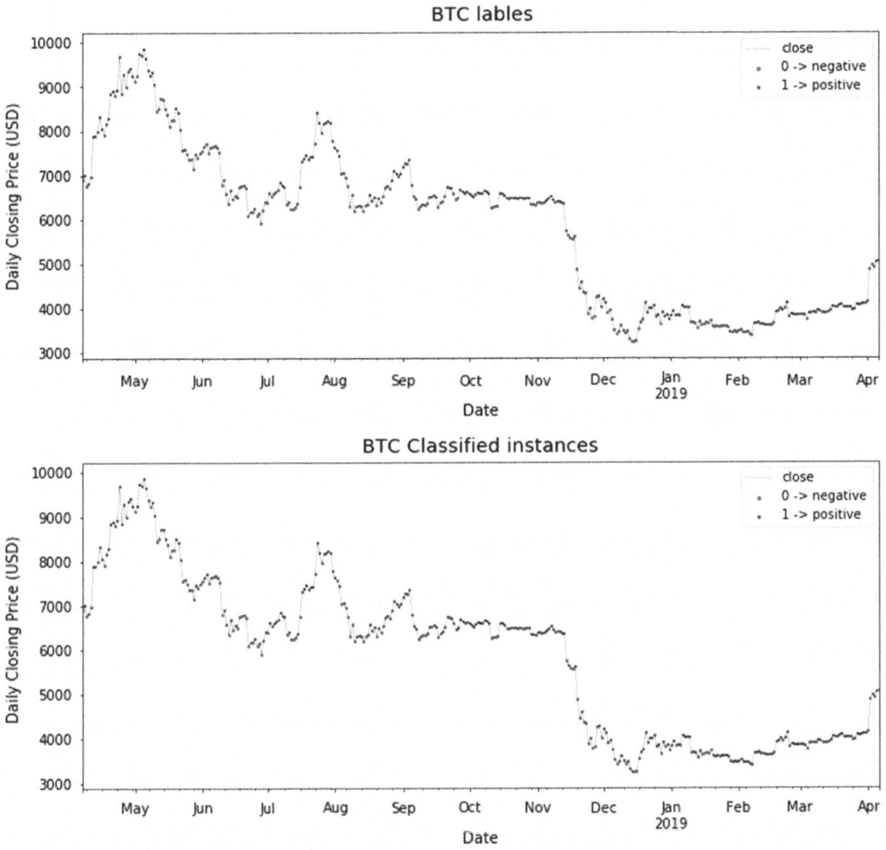

Fig. 7 Resulting labels for BTC, from 07/04/2018 to 06/04/2019, by an NN with four hidden layers (a fully-connected layer, a BRNN with LSTM layers, a LSTM layer and a fully-connected layer (Acc: 78%, Prec: 72%, Rec: 90%)

We apply the same methods for the different groups of features (Table 1). Again, as expected, results indicate that the most important scopes are the financial and blockchain ones (Table 6).

FNNs and RNNs with multiple coins. A similar NN to previous experiment, but with more nodes in each hidden layer, is trained for totally 73 coins. The NN consists of a fully-connected layer of 128 nodes, a BRNN with LSTM layers of 256 nodes, followed by another LSTM layer of 256 nodes and a final fully-connected layer with 128 nodes. We use the daily relevant change of each feature for the 73 coins as input. The features used are a subset of the financial and blockexplorer features. Namely, the features are: all the financial scope and the active addresses, exchange volume, fees, median fee, median transaction value, transaction count and transaction volume. The network was trained with time series of 10-days window and for test set we keep the last year for all the coins that is available. Thus, we get a dataset with 31546 train

Table 2 Results of FNN and RNN with multiple coins. Each coin is reported with the accuracy of the model on its last year's instances

Coin (Acc)
ADA (52.3%), AE (57.8%), AION (68.8%), ANT (68.5%), BAT (68.8%), BCH (67.7%), BNB (67.1%), **BTC (77.8%)**, BTG (60.3%), BTM (57.3%), **CENNZ (72.4%)**, CTXC (67.1%), CVC (56.7%), **DAI (83.6%)**, DASH (58.9%), DCR (61.4%), DGB (65.5%), DOGE (67.7%), DRGN (60.5%), ELF (63.0%), ENG (61.4%), EOS (64.7%), ETC (65.2%), ETH (59.5%), ETHOS (65.2%), FUN (59.2%), GAS (66.0%), GNO (61.1%), GNT (65.8%), GUSD (50.5%), ICN (57.3%), ICX (59.5%), KCS (45.8%), KNC (65.8%), LOOM (46.2%), LRC (54.8%), LSK (65.2%), LTC (67.7%), MAID (62.5%), **MANA (71.5%)**, MTL (62.7%), NAS (68.5%), NEO (54.5%), OMG (60.5%), **PAX (72.7%)**, PAY (55.6%), PIVX (57.5%), POLY (60.8%), POWR (58.1%), PPT (58.6%), QASH (59.7%), REP (68.2%), RHOC (66.6%), SALT (69.6%), SNT (61.6%), SRN (61.6%), TRX (59.5%), TUSD (54.5%), **USDC (84.6%)**, **USDT (89.0%)**, **VERI (72.1%)**, VET (56.2%), VTC (67.9%), WAVES (51.2%), WTC (61.9%), XEM (51.0%), XLM (69.0%), XMR (66.0%), **XRP (73.2%)**, XVG (44.1%), ZEC (55.1%), ZIL (63.6%), ZRX (65.2%)

and 25667 test instances. The results were Acc: 63%, Prec: 58%, Rec: 59%. Even though the overall accuracy is relatively low, it is very encouraging to notice that the performance on some specific coins (bold in Table 2) appears to be comparable to previous experiments.

3.2 *Regression*

In order to forecast price fluctuations we employ regression by NNs. The criteria are mean squared error (MSE) and mean absolute percentage error (MAPE) over n instances. MSE equals the average squared difference between predicted F_t and true A_t values for $t = 1, \ldots, n$, while MAPE measures the percentage (relative) difference between predicted and true values, as follows.

$$\text{MSE} = \frac{1}{n}\sum_{t=1}^{n}(A_t - F_t)^2, \qquad \text{MAPE} = \frac{100\%}{n}\sum_{t=1}^{n}\left|\frac{A_t - F_t}{A_t}\right|.$$

RNN. We design an NN with two hidden layers of 128 nodes. The first layer is an LSTM with layer normalization [3] and recurrent dropout [20] followed by a simple LSTM layer. The former normalizes layer output and is quite effective at stabilizing the hidden state dynamics. The recurrent dropout is a technique used to avoid overfitting and improve results by dropping some neurons using a prescribed keep-probability. Here we train the NN, with 80% keep probability for the dropout layer, using BTC's financial scope (OHLCV and Market Cap) and the results of an Autoregressive integrated moving average (ARIMA) model (total of 7 features for each instance). ARIMA are commonly used models for time series forecasting and their results here are used to improve the fluctuation forecasting. The parameters of the ARIMA model were selected using an auto ARIMA implementation (lag order: 5, degree of differencing: 1, order of moving average model: 0).

Training uses a 3-day time-step window per instance, using BTC prices from 01/01/2015 to 04/01/2018. The test set contains BTC prices from 05/01/2018 to 27/01/2019, with total MSE: 0.00119, MAPE: 7.28%. In Fig. 8 we observe the similarity of forecast against the actual daily closing price fluctuation.

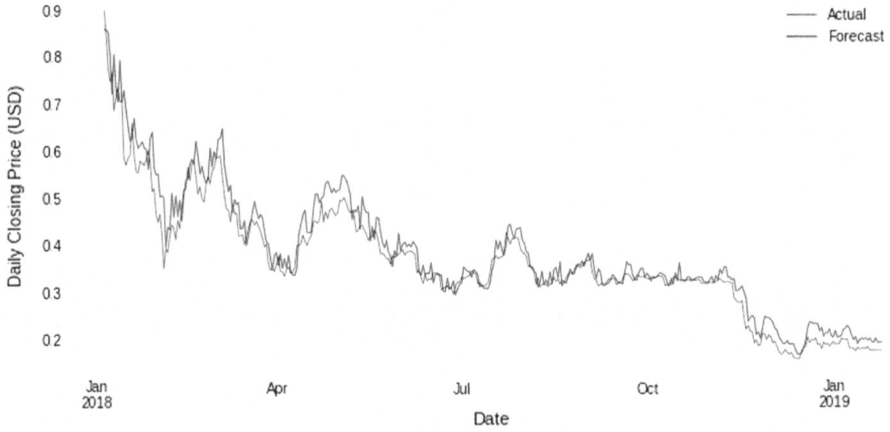

Fig. 8 BTC price forecasting from 05/01/2018 to 27/01/2019 (red) versus actual closing price normalized (green) (color figure online)

Table 3 Results summary for models

NN	Train	Test	Train time (s)	Test time (s)	Accuracy (%)
FNN	BTC	BTC	10.4	0.65	63.6
GRU	BTC	BTC	8.51	0.75	71.0
BRNN	BTC	BTC	9.85	0.81	72.0
LSTM	BTC+LTC+ETH	DASH	80.0	0.81	64.0

Table 4 Results summary for models with the daily difference of each feature (from previous day) as input

NN	Train	Test	Train time (s)	Test time (s)	Accuracy (%)
FNN	BTC	BTC	9.3	0.63	70.3
FNN + RNN	BTC	BTC	194	0.48	78
FNN + RNN	73 coins	73 coins	1300	24	63

Table 5 Feature importance. Results are sorted by permutation importance MSE

	Permutation importance			Repeating value importance		
Feature	Acc. (%)	MAPE (%)	MSE	Acc. (%)	MAPE (%)	MSE
Realized cap	70.8	8.9	0.005397	76.2	2.1	0.000270
Close	74.3	4.5	0.001853	74.8	3.9	0.000908
Low	79.3	3.3	0.001118	79.7	2.5	0.000368
Exchange volume	77.2	3.7	0.001031	78.1	0.4	0.000008
Market cap	78.2	3.4	0.000868	77.5	0.4	0.000008
Average difficulty	80.2	3.1	0.000688	78.9	1.4	0.000120
Transaction count	76.4	2.9	0.000659	77.8	0.0	0.000000
Payment count	80.1	3.0	0.000561	80.0	2.8	0.000480
Adjusted transaction volume	79.6	2.6	0.000542	79.2	1.8	0.000188
High	75.7	2.7	0.000525	75.3	3.2	0.000608
Open	77.2	2.2	0.000524	78.4	0.7	0.000030
Active addresses	76.7	2.4	0.000426	77.3	0.7	0.000030
Fees	79.5	2.1	0.000315	80.5	3.5	0.000751
Block size	76.4	1.8	0.000246	75.9	2.5	0.000368
Pull requests merged	77.8	1.7	0.000228	79.5	2.1	0.000270
Total issues	77.9	1.1	0.000195	77.8	0.0	0.000000
Forks	79.1	1.6	0.000179	78.4	0.7	0.000030
Pull request contributors	78.8	1.3	0.000167	78.4	0.7	0.000030
Commit count (4 weeks)	78.5	1.3	0.000161	79.5	2.1	0.000270
Block count	78.1	1.3	0.000161	78.4	0.7	0.000030
Stars	78.7	1.2	0.000101	79.5	2.1	0.000270
Subscribers	78.6	1.1	0.000096	78.6	1.1	0.000068
Transaction volume	77.5	1.1	0.000092	76.4	1.8	0.000188
Volume	77.5	0.8	0.000075	78.9	1.4	0.000120
Generated coins	78.3	0.6	0.000050	77.0	1.1	0.000068
Median fee	78.0	0.6	0.000029	78.4	0.7	0.000030
Closed issues	78.0	0.2	0.000008	77.8	0.0	0.000000
Median transaction value	77.8	0.1	0.000002	77.8	0.0	0.000000

Table 6 Features importance per scope

	Permutation importance			Repeating value importance		
Scope	Acc. (%)	MAPE (%)	MSE	Acc. (%)	MAPE (%)	MSE
Financial	57.9	25.6	0.040344	54.0	30.6	0.056814
Blockchain explorers	72.2	7.2	0.003561	75.1	3.5	0.000751
Developer (Git)	78.5	2.0	0.000288	81.9	5.3	0.000751

4 Discussion and Future Work

We have presented a comparative analysis of NNs to classify cryptocurrencies and forecast their price fluctuations. We have included several features to cover most of their aspects and evaluated their relevance. Besides daily prices (open, high, low, close prices, volume and market cap) and blockchain features (number of transactions, blocks, active addresses, fees, etc), we have also used features to describe the development and software code popularity and penetration into the community (stars, subscribers, forks, commit counts, etc).

The classification of each instance as positive or negative, based on the daily change of features relative to their previous value, seems to provide good results, since the accuracy on last year's instances is 78% by a NN with a fully-connected layer of 64 nodes, a BRNN with LSTM layers of 128 nodes, followed by a LSTM layer of 128 nodes and a fully-connected layer with 64 nodes. Positive and negative instances are those where daily closing prices are increasing or decreasing respectively, compared to the previous days. Therefore, its direct purpose is not to predict, but to evaluate the current snapshot of a coin. Prototypes based on similar ideas are provided by the first two authors on GitHub and daily results are reported on the VESPUCCI website.[2]

Concerning prediction, the regression results of an RNN with two LSTM cells of 128 nodes, are quite promising in forecasting price fluctuations with total error of 7.28% from actual price, using only daily prices and an ARIMA prediction. Integrating the remaining features to a more complex architecture should provide more accurate forecasts.

Even though the NNs used were fed with raw features from different scopes, results already appear to be promising. New features shall be considered e.g. technical indicators like moving average convergence divergence (MACD), relative strength index (RSI) as well as sentiment analysis. The combination of features from various widely traded cryptocurrencies to generate a larger dataset, would be suitable for deep learning methods. More powerful, deep neural network architectures shall be evaluated. Using several layers, and state-of-the-art deep learning, should enhance forecasting power. For this, Autoencoders and Generative Adversarial Networks (GANs) shall be considered.

[2]https://github.com/PythagorasdotSystems/, https://vespucci.site/.

Acknowledgements The authors thank Alexis Anastasiou of Pythagoras Systems and Jason Theodorakopoulos for their precious comments and suggestions throughout this work. The first two authors are partially supported by the European Union's H2020 research and innovation programme under grant agreement No 734242 (LAMBDA). The first author was also partially employed by Pythagoras Systems during this work.

References

1. Abadi, M., Barham, P., Chen, J., Chen, Z., et al.: Tensorflow: a system for large-scale machine learning. In: Proceedings of the USENIX Symposium on Operating Systems Design & Implement, pp. 265–283 (2016)
2. Ariyo, A., Adewumi, A., Ayo, C.: Stock price prediction using the ARIMA model. In: Proceedings of the UKSim-AMSS International Conference on Computer Modelling and Simulation, pp. 106–112 (2014)
3. Ba, J., Kiros, R., Hinton, G.E.: Layer normalization. Tech. Rep. 1607, arXiv (2016)
4. Calès, L., Chalkis, A., Emiris, I.Z., Fisikopoulos, V.: Practical volume computation of structured convex bodies, and an application to modeling portfolio dependencies and financial crises. In: Proceedings of the International Symposium on Computational Geometry, Budapest, pp. 19:1–19:15 (2018)
5. Chatzis, S.P., Siakoulis, V., Petropoulos, A., Stavroulakis, E., Vlachogiannakis, N.: Forecasting stock market crisis events using deep and statistical machine learning techniques. Expert. Syst. Appl. **112**, 353–371 (2018)
6. Cho, K., van Merrinboer, B., Gulcehre, C., Bougares, F., et al.: Learning phrase representations using rnn encoder-decoder for statistical machine translation (2014). https://doi.org/10.3115/v1/D14-1179
7. Chung, J., Gulcehre, C., Cho, K., Bengio, Y.: Empirical evaluation of gated recurrent neural networks on sequence modeling (2014)
8. Hileman, G., Rauchs, M.: 2017 global cryptocurrency benchmarking study. SSRN Electron. J. (2017)
9. Hochreiter, S., Schmidhuber, J.: Long short-term memory. Neural Comput. **9**, 1735–1780 (1997). https://doi.org/10.1162/neco.1997.9.8.1735
10. Jang, H., Lee, J.: An empirical study on modeling and prediction of bitcoin prices with bayesian neural networks based on blockchain information. IEEE Access **6**, 5427–5437 (2018)
11. Jiang, Z., Liang, J.: Cryptocurrency portfolio management with deep reinforcement learning (2017). https://doi.org/10.1109/IntelliSys.2017.8324237
12. Kim, Y.B., Lee, J., Park, N., Choo, J., et al.: When Bitcoin encounters information in an online forum: using text mining to analyse user opinions and predict value fluctuation. PLOS One **12**, e0177630 (2017)
13. Kristjanpoller, W., Minutolo, M.C.: A hybrid volatility forecasting framework integrating GARCH, artificial neural network, technical analysis and principal components analysis. Expert. Syst. Appl. **109**, 1–11 (2018)
14. Lahmiri, S., Bekiros, S.: Cryptocurrency forecasting with deep learning chaotic neural networks. Chaos Solitons Fractals **118**, 35–40 (2019)
15. McNally, S., Roche, J., Caton, S.: Predicting the price of bitcoin using machine learning. In: Euromicro International Conference on Parallel, Distributed & Network-Based Processing (PDP), pp. 339–343 (2018)
16. Nakamoto, S.: Bitcoin: a peer-to-peer electronic cash system. Cryptography Mailing list at https://metzdowd.com (2009)
17. Nakano, M., Takahashi, A., Takahashi, S.: Bitcoin technical trading with artificial neural network. Phys. A Stat. Mech. Appl. **510**, 587–609 (2018)

18. Saad, M., Mohaisen, A.: Towards characterizing blockchain-based cryptocurrencies for highly-accurate predictions. In: Proceedings of the IEEE Conference on Computer Communications (Infocom) Workshops, pp. 704–709 (2018)
19. Schuster, M., Paliwal, K.: Bidirectional recurrent neural networks. IEEE Trans. Signal Process. **45**, 2673–2681 (1997). https://doi.org/10.1109/78.650093
20. Semeniuta, S., Severyn, A., Barth, E.: Recurrent dropout without memory loss. In: Proceedings of the COLING International Conference on Computational Linguistics, pp. 1757–1766 (2016)

Decentralized Incentive-Compatible and Sybil-Proof Transaction Advertisement

Oğuzhan Ersoy, Zekeriya Erkin and Reginald L. Lagendijk

Abstract In a blockchain network, transaction advertisement is the announcement of the new transactions to the participants (miners) who are responsible to validate them. Existing blockchain protocols lack an incentive-compatible advertisement process where a rational participant would gain from advertising a transaction. The deficiency can be solved by a Sybil-proof rewarding function which divides the transaction fee among the round leader and the nodes who advertise it. Up to now, there have been three rewarding function proposals, all of which require special constraints on the blockchain network model, e.g., tree-structured connections. In this work, we formulate the rewarding function and obtain the necessary conditions for Sybil-proofness and incentive-compatibility properties. To the best of our knowledge, we present the first rewarding function which is suitable for any blockchain network model. We introduce path length dependent rewarding for the nodes involved in the advertisement process, which helps us to overcome the impossibility results given in the previous works. Our rewarding function divides the transaction fee among the nodes who advertise it, the current round leader and the next round leader. In addition to these achievements, unlike previous proposals, our rewarding function provides resistance against the forking attacks where an adversary rejects a valid block and creates a fork to gain the transaction fees in the original block.

Keywords Blockchain · Transaction advertisement · Incentive mechanism

O. Ersoy (✉) · Z. Erkin · R. L. Lagendijk
Cyber Security Group, Department of Intelligent Systems,
Delft University of Technology, Delft, The Netherlands
e-mail: o.ersoy@tudelft.nl

Z. Erkin
e-mail: z.erkin@tudelft.nl

R. L. Lagendijk
e-mail: r.l.lagendijk@tudelft.nl

P. Pardalos et al. (eds.), *Mathematical Research for Blockchain Economy*,
Springer Proceedings in Business and Economics,
https://doi.org/10.1007/978-3-030-37110-4_11

1 Introduction

A blockchain is commonly defined as a decentralized transaction ledger which is updated by a peer-to-peer structured network of parties. Update of the ledger can be accumulated into two main functionalities: advertisement and chain extension. Each party becomes aware of the newly made transactions during the advertisement and, in the chain extension, a round leader adds these advertised transactions into a block which extends the chain.

Assuming parties involved in the blockchain ecosystem are rational where they try to maximize their profit, each functionality of the blockchain protocol must be incentive compatible [16, 18]. Blockchain protocols have two types of incentives: block rewards and transaction fees. Block reward is received for each block and the reward is fixed by the consensus protocol, whereas each transaction fee is given to the round leader who adds the corresponding transaction and the fee determined by the sender of the transaction. It can be seen that the existing protocols including Bitcoin [14] and Ethereum [19] rewards only the round leader, i.e., the chain extension functionality. As a result, the protocols supposed to operate on a decentralized or peer-to-peer manner controlled by centralized pools [15], e.g., Bitcoin and Ethereum mining pools [9]. Moreover, transaction advertisement is altruistically fulfilled [17], which is not sustainable in the long-term [10, 12].

Incentive-compatible advertisement process can be done with a rewarding function which divides the transaction fee among the round leader and the nodes who advertise it. There have been three proposals for the rewarding function [1, 2, 6]. In [2], Babaioff et al. presented the first game theoretical analysis of the transaction advertisement functionality of the Bitcoin and they proposed a solution for d-ary directed tree networks. In [1], Abraham et al. presented a rewarding function to propagate transactions and validated blocks (puzzles) where the propagation requires the parties possessing the transaction to have common neighbors in the network. Recently, Ersoy et al. [6] formalized the rewarding functionality and proposed a solution for well-connected networks. Under the model description in [6], they also proved that it is impossible to have a Sybil-proof rewarding function for poorly-connected networks. All in all, it can be said that none of the existing rewarding functions is Sybil-proof for the generic network model.

1.1 Our Contributions

In this work, to the best of our knowledge, we present the first incentive-compatible and Sybil-proof rewarding function which works for any network model. We introduce path-length dependent rewarding for the advertising nodes and the round leader which helps us to break the impossibility results given in [6]. Our rewarding function divides the transaction fee among the nodes who advertise it, the current round leader and the next round leader with respect to the following formula:

$$r_\ell^{[i]}: \begin{cases} F \cdot \sum_{j=i}^{\ell} \left((-1)^{j-i} \cdot \binom{\ell-i}{j-i} \cdot \prod_{k=1}^{j-1}(1-\Gamma_k)\right) & \text{for } 1 \le i < \ell, \\ F \cdot \prod_{k=1}^{\ell-1}(1-\Gamma_k) & \text{for } i = \ell. \end{cases}$$

where F is the fee, $r_\ell^{[i]}$ is the reward of the ith node in the advertisement path, $r_\ell^{[NRL]} = F - \sum_{i=1}^{\ell} r_\ell^{[i]}$ is the reward of the next round leader, and Γ_k's are the network variables. The ingenuity behind this formula is the equality $r_\ell^{[i]} = r_{\ell+1}^{[i]} + r_{\ell+1}^{[i+1]}$ satisfied for all ℓ and i values, which discourages the nodes from introducing Sybil nodes.

In our rewarding function, round leaders benefit from the reward of the previous blocks which discourages forking attacks. In [3], it is shown that (high) transaction fees would yield *forking attacks* where the malicious party does not accept the block with high transaction fees mined by others and tries to fork the chain by adding these transactions to his block. In order to mitigate these forking attacks, transaction fees can be shared with the round leader of the next round [7]. In other words, the countermeasure of forking attacks utilizes a rewarding mechanism where the round leaders are rewarded from the previous blocks. Our rewarding function, unlike previous works, is compatible with these countermeasures since the next round leader also benefits from the transaction fees collected in the previous block.

The rest of the paper is organized as follows: Sect. 2 introduces the notation and formulates the rewarding function for the transaction advertisement process. Section 3 presents detailed related work. Section 4 presents our rewarding function. Finally, Sect. 5 concludes the paper.

2 Model

In this section, we introduce our model and terminology used in this paper. Then, we define the rewarding function which provides the incentive for the advertisement process.

Network. We model the blockchain network as a graph where each party represented with a *node*. For the rest of the paper, we will refer to each party as a node in the graph. The identity of a node is its public key for the corresponding PKI defined in the blockchain protocol, which is beyond the scope of this paper. We do not assume a link between IP address of a node with its public key.

Nodes. In general, there are two types of nodes involved in a blockchain protocol: *clients* and *miners*. Clients make new transactions and check their validity by reading (some part of) the chain, whereas miners are responsible to operate the blockchain by validating transactions and creating blocks. For simplicity reasons, in our model, nodes are not restricted to specific roles, i.e., every node can be a client or a miner at the same time.

Mining. In our model, anyone can join the blockchain network and become a miner which is the case for the *permissionless* blockchains. $\gamma(n_i)$ denotes the capacity of miner (node) n_i, i.e., the probability of being the round leader. For the proof-of-work (or proof-of-stake) based blockchains, capacity corresponds to the proportional CPU power (or stake value) over the rest of the network.

Sybil nodes. We adopt the definition of a *Sybil node* given in [2]: fake nodes with new identities having the same neighbors with the original node. Two nodes with different identities (public keys) controlled by the same entity can be seen as a joint node where either one of them can be seen as a Sybil node. On the other hand, every individual node contributing to the connectivity of the network is assumed to be not Sybil. Thus, for example, client nodes are not considered as Sybil nodes though they are not involved in the mining process.

Regarding the advertisement process of a transaction, the corresponding client is the source and round leader is the destination of the advertisement. Depending on the consensus algorithm used in the blockchain protocol, the round leader may or may not be publicly known before the advertisement process. $\mathcal{N}^T$ denotes the set of nodes who have received the transaction T, whereas $\overline{\mathcal{N}^T}$ is set of nodes who did not receive transaction T yet.

2.1 Rewarding Mechanism

A blockchain network consists of rational nodes (miners) which can create Sybil nodes. In this manner, each functionality should be Sybil-proof and incentive-compatible [16]. Regarding the transaction advertisement functionality, transaction fees are the only incentive instruments. Our aim is to determine the ideal way of dividing the fee among the advertising nodes and the round leader(s). Definition 1 presents a modified version of the rewarding function defined in [6]. The same definition is implicitly used in the other works as well [1, 2].

Definition 1 (*Rewarding Function*). Rewarding function $\mathcal{RF}$ divides the fee of a transaction among the advertising (propagating) nodes and the round leader(s). Let $\mathbf{P}$ be the path of a transaction advertised from the client to the round leader and F is the fee defined by the client, then $\mathcal{RF}$ can be formulated as

$$\mathcal{RF} : \{F, \mathbf{P}\} \longrightarrow \{r_\ell^{[i]}\}_{i=1}^{\ell} : \quad \sum_{i=1}^{\ell} r_\ell^{[i]} = R_\ell \leq F, \; r_\ell^{[NRL]} = F - R_\ell, \qquad (1)$$

where length of the path is $|\mathbf{P}| = \ell$ and $r_\ell^{[i]} > 0$ is the reward of the ith node in the path $\mathbf{P}$. R_ℓ denotes the total reward given to the nodes in the path $\mathbf{P}$ and the remaining part of the fee F is described as the reward of the next round leader denoted by $r_\ell^{[NRL]}$.

Here, $r_\ell^{[\ell]}$ denotes the reward of the round leader as the last node in the path. In other words, the round leader is considered as a part of the advertisement path.

Yet, we may mention the round leader additional to the nodes in the path for a clear separation of the roles.

The main difference between our definition in Definition 1 and the one used in previous works is that the total reward of the nodes in a path may vary with respect to the length of the path. Whereas, in the previous works, the total reward of the advertising nodes and the round leader is equal to the transaction fee and $r_\ell^{[NRL]}$ is equal to zero. This flexibility enables us to break the impossibility results given in [6]. In addition, rewarding the next round leader ($r_\ell^{[NRL]}$) helps to mitigate the forking attacks [7].

The following definition describes *ideal rewarding functionality* with respect to the properties determined in the previous works.

Definition 2 (*Ideal Rewarding Function*). An *ideal* rewarding function satisfies the following properties:

1. Sybil-proofness: No node would benefit from introducing Sybil nodes.
2. Incentive-compatibility: The nodes who are aware of a transaction benefit from advertising it.

The incentive-compatibility is defined with respect to the individual rationality constraint where the nodes try to maximize their profits. We do not take into account the cases where the nodes have other incentives than maximizing their profits, e.g., trying to minimize the utility of the rival.

Rewarding the path. The rewarding function divides the fee among the nodes who are in the advertisement path and the next round leader. It is important to note that the client does not choose a specific advertisement path in advance. Because of the nature of the peer-to-peer propagation mechanisms, the same transaction is very likely to be propagated (advertised) via several paths. For each transaction, the round leader decides to add the path he prefers, which maximizes his profit. Then, the reward is divided between the node who are recorded on the path and the next round leader. For all the other paths which are not chosen by the round leader, there will be no rewarding.

Immutability of the path. The advertisement paths should be immutable because an intermediary node may try to subtract some of the previous nodes in the path to increase its profit. Similarly, the round leader may try to alter the path. The existing rewarding mechanisms suggest to use a chain of signatures where each sender node suffixes the public key of the receiving node and signs it before propagating the transaction. Therefore, each path can be validated by tracking back to the client node and it is not possible to add or remove a node from the path. In this work, we utilize the same mechanism to prevent any kind of manipulation in the path. Details of the mechanism can be found in [1].

Generation of the path. During the advertisement of a transaction, the transaction flow from the client to the nodes generates a graph model. The model plays an important role to determine the action of a rational node. In an advertisement process, three types of graph models can be observed (see Fig. 1):

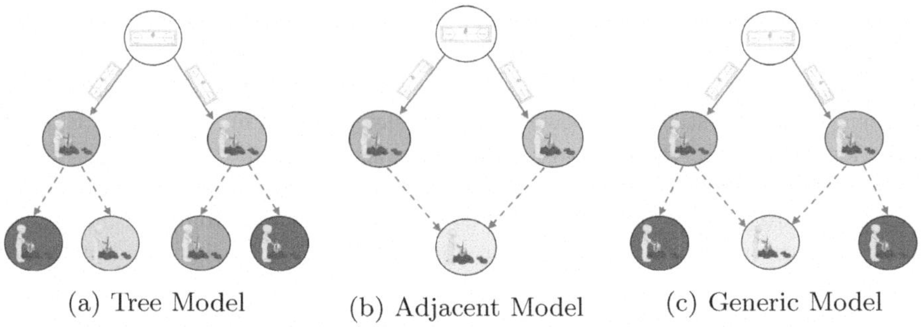

Fig. 1 Graph models for the blockchain network. For each model, the top node represents the client and the rest is the flow of the transaction

- *Tree model* where the transaction flow from the client to the nodes is assumed to create a tree structure. Regarding the advertisement, there is no competition between nodes for their individual subsequent (or child) nodes.
- *Adjacent model* where the nodes advertised about the transaction are assumed to have common neighbors. Thereby, there is always advertisement competition for the subsequent nodes.
- *Generic model* where the transaction flow may create any type of graph and the nodes may or may not compete for advertising the subsequent nodes.

In this work, we do not have any restriction on the graph model of the blockchain network, and thereby we assume the generic model.

3 Related Work

The lack of incentive for the transaction advertisement has been studied in peer-to-peer networks for several scenarios [4, 5, 11, 13]. Yet, the proposed solutions are not applicable to a blockchain protocol for the following reasons. First, the transactions in [11, 13] have a specific destination, whereas transactions in a blockchain network are sent to the round leader and anyone is a potential round leader. Therefore, there is also a competition between nodes who advertised the transaction and the subsequent nodes. Second, the other works [4, 5] analyze a multi-level marketing scenario where every node advertising the transaction is rewarded if they succeed to a purchase. However, in a blockchain network, only the parties who are recorded on the advertisement path are rewarded. Third, because of the transaction fees in a blockchain, rational nodes have an incentive not to propagate the transaction, whereas there is no benefit of non-propagation in [4, 5, 11, 13]. Consequently, transaction advertisement mechanisms proposed for peer-to-peer networks are not compatible with blockchain protocols. Recently, there have been proposals on incentive-compatible advertisement mechanisms dedicated for the blockchain protocols [1, 2, 6].

The first work on the incentive-compatibility of the transaction advertisement for a blockchain network is done by Babaioff et al. in [2]. They discovered that a rational node (miner) in Bitcoin has no incentive to propagate a transaction. They analyzed a specific type of graph model for the blockchain network, which is regular d-ary directed tree with height H (see Fig. 1). It is also assumed that nodes have the same capacity. Under these assumptions, they proposed a Sybil-proof and incentive-compatible rewarding mechanism which survives iterated removal of dominated strategies. Moreover, the authors show that there is no Sybil-proof rewarding mechanism which constitutes the dominant strategy for all nodes.

Abraham et al. [1] proposed a consensus mechanism, Solidus, offering an incentive to propagate transactions and blocks. In their proposal, the amount of transaction fee passed to the next node is determined by the sender. Their work is based on the adjacent network model (see Fig. 1) where the nodes having the transaction (must) have a common neighboring node. They concluded that each node should charge a fixed amount of the fee and pass the rest to the next node. Their proposal requires a detailed study on determining the charging amount since a small charging value would not lead to propagation while a high value may not have enough incentive for the round leader.

Ersoy et al. [6] recently published an incentive mechanism for the transaction advertisement. They analyzed the generic network model (see Fig. 1) where the participants having the transaction may or may not have common neighbors. They proved that there is no Sybil-proof rewarding mechanism for 1-connected networks where removal of a party would disconnect the network, which extends the impossibility result given in [2]. They present a Sybil-proof and incentive-compatible rewarding function suitable for any 2- or more connected networks.

4 Our Rewarding Function

In this section, we construct the ideal rewarding function with respect to Definition 2 by formalizing the Sybil-proofness and incentive-compatibility properties.

The rewarding function is used to divide the transaction fee F among the nodes who advertised (propagated) it, the round leader and the next round leader. The total reward given to the nodes in the path (including the round leader) does not have to be equal to the fee determined by the client. In other words, in Eq. (1), the reward of a path of length ℓ is upper bounded by the fee, i.e., $R_\ell \leq F$. The difference between the fee and total reward of the path is given to the next round leader, i.e., $r_\ell^{[NRL]} = F - R_\ell$.

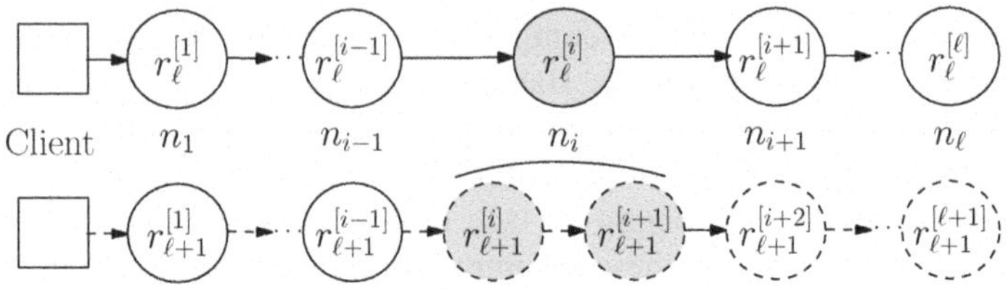

Fig. 2 Illustration of a Sybil node added by node n_i. The rewards of nodes are presented for both cases

4.1 Sybil-Proofness

According to [2], a Sybil node can be defined as a *fake* node attached to the original one and has the same connections with the original node. Figure 2 illustrates a Sybil node created by a node n_i. In order to prevent n_i from introducing a Sybil node, the reward of n_i must decrease (or stay the same) after introducing the Sybil node, which can be formulated as $r_\ell^{[i]} \geq r_{\ell+1}^{[i]} + r_{\ell+1}^{[i+1]}$. By generalizing this condition for any path length and any position in a path, the Sybil-proofness condition can be formulated as

$$\forall \ell, \forall i \in \{1, \ldots, \ell\} \;\; r_\ell^{[i]} \geq r_{\ell+1}^{[i]} + r_{\ell+1}^{[i+1]}. \tag{2}$$

Lemma 1 *The total reward of path decreases as the path length increases expect for the length of 1, i.e.,*

$$R_1 \geq R_2, \;\; \forall \ell \geq 2: \;\; R_\ell > R_{\ell+1}. \tag{3}$$

Proof Using Eq. (2), for any ℓ, by summing all possible $i \in \{1, \ldots, \ell\}$ values:

$$\sum_{i=1}^{k} r_k^{[i]} \geq \sum_{i=1}^{k} r_{k+1}^{[i]} + \sum_{i=1}^{k} r_{k+1}^{[i+1]} \implies R_\ell \geq R_{\ell+1} + R_{\ell+1} - r_{\ell+1}^{[1]} - r_{\ell+1}^{[\ell+1]}. \tag{4}$$

Note that $R_{\ell+1} - r_{\ell+1}^{[1]} - r_{\ell+1}^{[\ell+1]}$ is greater than or equal to zero and the equality holds for $\ell = 1$. For the rest, it is strictly greater than zero since $r_{\ell+1}^{[i]} > 0$ $(\forall \ell > i > 1)$. □

Remark 1 Lemma 1 coincides with the impossibility result given in [6] where the authors assumed that the total reward is always equal to the fee. As a result of the Eqs. (1) and (3), the reward of the next round leader increases as the path length increases, i.e., $r_\ell^{[NRL]} < r_{\ell+1}^{[NRL]}$ for $\ell \geq 2$. Since the previous models in [1, 2, 6] implicitly assume that $r_\ell^{[NRL]}$ is equal to zero for all ℓ values, these models could not provide Sybil-proofness for a single advertisement path.

Corollary 1 *The total path reward of length* ℓ*,* R_ℓ*, can be upper bounded by the following inequality* $F \geq R_1 \geq R_2 \geq R_\ell + \sum_{j=3}^{\ell+1} \sum_{i=2}^{j-1} r_j^{[i]}$.

The corollary can be proved by using the Eq. (4). In order to decelerate the decrease in the total path reward R_ℓ while preserving the Sybil-proofness, we fix the Sybil-proofness condition as

$$\forall \ell, \forall i \in \{1, \ldots, \ell\} \;\; r_\ell^{[i]} = r_{\ell+1}^{[i]} + r_{\ell+1}^{[i+1]}, \tag{5}$$

which results in $R = R_1 = R_2 = R_\ell + \sum_{j=3}^{\ell+1} \sum_{i=2}^{j-1} r_j^{[i]}$. Equation (5) enables computation of the rewards of intermediary nodes with respect to the (potential) rewards for the round leaders.

Lemma 2 *The reward of the* $i-th$ *node in the path of length* ℓ *is equal to*

$$r_\ell^{[i]} = \sum_{j=i}^{\ell} (-1)^{j-i} \cdot \binom{\ell-i}{j-i} \cdot r_j^{[j]}. \tag{6}$$

Proof (Proof by induction). We have two inductions on ℓ and i respectively. We start with an induction on ℓ.

Base case ($\ell = 1, 2$): For the first values of ℓ and i, the equality is satisfied since $r_1^{[1]} = r_1^{[1]}$, $r_2^{[1]} = r_1^{[1]} - r_2^{[2]}$ and $r_2^{[2]} = r_2^{[2]}$ from Eq. (5).
Hypothesis ($\ell > \ell'$): Assume that Eq. (6) holds for all ℓ' values less than ℓ.
Inductive step for ℓ: Now, we show that Eq. (6) holds for ℓ.

For the fixed value of ℓ, we have another induction on i with decreasing order.
Base case ($i = \ell$): For $i = \ell$, the equality is satisfied since $r_\ell^{[\ell]} = r_\ell^{[\ell]}$, $r_\ell^{[\ell-1]} = r_{\ell-1}^{[\ell-1]} - r_\ell^{[\ell]}$ from Eq. (5).
Hypothesis ($i < i'$): Assume that Eq. (6) holds for all i' values greater than i.
Inductive step for i: Here, we show that Eq. (6) holds for i. Using the Eq. (5) and the hypothesis:

$$\begin{aligned}
r_\ell^{[i]} &= r_{\ell-1}^{[i]} - r_\ell^{[i+1]} \\
&= \sum_{j=i}^{\ell-1} (-1)^{j-i} \cdot \binom{\ell-i-1}{j-i} \cdot r_j^{[j]} - \sum_{j=i+1}^{\ell} (-1)^{j-i-1} \cdot \binom{\ell-i}{j-i-1} \cdot r_j^{[j]} \\
&= r_i^{[i]} + \sum_{j=i+1}^{\ell-1} (-1)^{j-i} \cdot \left(\binom{\ell-i-1}{j-i} - \binom{\ell-i}{j-i-1}\right) \cdot r_j^{[j]} + r_\ell^{[\ell]} \cdot (-1)^{l-j} \\
&= \sum_{j=i}^{\ell} (-1)^{j-i} \cdot \binom{\ell-i}{j-i} \cdot r_j^{[j]},
\end{aligned}$$

which proves the inductive step for i and thereby the step for ℓ. □

4.2 Incentive Compatibility

Under the rational behavior assumption where each node acts to maximize its profit, the advertisement of transactions should be profitable for the nodes. A rational node would take the action of advertisement for a transaction if it is more profitable or at least not less than before. In other words, the expected reward of a node after the advertisement should be greater than or equal to the reward before the action.

For a transaction T, at the beginning, we can assume that some of the nodes are aware of T because the client would advertise to its neighbors. From that point, we analyze the necessary incentives to advertise T to the rest of the network. In that sense, advertising the transaction to the nodes who already have it is irrelevant for our analysis. Therefore, advertising implicitly refers to the ones who do not have it yet. From a node's point of view, a neighbor node who has the transaction ($\in \mathcal{N}^T$) can be seen the other way around ($\in \overline{\mathcal{N}^T}$). These illusory views may lead to advertising T to a node who already has it. Yet, they do not obstruct our advertisement purpose since a node in $\overline{\mathcal{N}^T}$ will not be seen as in $\mathcal{N}^T$.

Lemma 3 *Let n be a rational node in $\mathcal{N}^T$ with distance ℓ to the client. n will advertise T to its neighbors who are not aware of it if the following inequality holds:*

$$r_{\ell+1}^{[\ell]} \geq r_{\ell}^{[\ell]} \cdot \frac{\gamma(n)}{\gamma(\mathcal{N}^T)}.$$

Proof Since n is a rational node, it would decide on an action which maximizes its expected reward from the transaction T. Let $\mathcal{A}_n \in \overline{\mathcal{N}^T}$ be the union of neighbors of n who do not have T. Now, we compute the expected reward of n for both actions:

- *Not advertising*: If n decides to not advertise T, the only way it would earn from T is being the round leader. Then, the expected reward of n can be computed as:

$$ER(n, T, act: not) = r_{\ell}^{[\ell]} \cdot \frac{\gamma(n)}{\gamma(\mathcal{N}^T)}, \tag{7}$$

  where $\frac{\gamma(n)}{\gamma(\mathcal{N}^T)}$ is the proportional probability of n being the round leader among $\mathcal{N}^T$. Here, the capacity of the $\overline{\mathcal{N}^T}$ is irrelevant because expected reward of $\overline{\mathcal{N}^T}$ from T is equal to zero. In other words, $\overline{\mathcal{N}^T}$ cannot earn the reward of T since it is not aware of T.
- *Advertising*: If n decides to advertise T, there are two possibilities n can earn from T: being the round leader or being an intermediary node for the paths advertised to $\mathcal{A}_n$. Then, the expected reward of n can be computed as:

$$ER(n, T, act: adv) = \frac{r_{\ell}^{[\ell]} \cdot \gamma(n) + r_{\ell+1}^{[\ell]} \cdot \gamma(\mathcal{A}_n)}{\gamma(\mathcal{N}^T) + \gamma(\mathcal{A}_n)}. \tag{8}$$

Node n would advertise T to $\mathcal{A}_n$ if $ER(n, T, act : adv) \geq ER(n, T, act : not)$. Using Eqs. (7) and (8), the advertisement decision can be formulated as:

$$\frac{r_\ell^{[\ell]} \cdot \gamma(n) + r_{\ell+1}^{[\ell]} \cdot \gamma(\mathcal{A}_n)}{\gamma(\mathcal{N}^T) + \gamma(\mathcal{A}_n)} \geq r_\ell^{[\ell]} \cdot \frac{\gamma(n)}{\gamma(\mathcal{N}^T)} \iff r_{\ell+1}^{[\ell]} \geq r_\ell^{[\ell]} \cdot \frac{\gamma(n)}{\gamma(\mathcal{N}^T)}. \quad (9)$$

□

Remark 2 From Lemma 3, specifically Eq. (9), it can be seen that the advertisement decision of a rational node is independent of the capacity of the receiving nodes. It solely depends on the node's proportional capacity with respect to the others who have the transaction.

Corollary 2 *Let $r_{\ell+1}^{[\ell]} = \Gamma_\ell \cdot r_\ell^{[\ell]}$ for some constant $\Gamma_\ell \in (0, 1)$. For a transaction T, any rational node n, with distance ℓ to the client, would advertise T as long as its proportional capacity with respect to the rest of $\mathcal{N}^T$ is less than or equal to Γ_ℓ, i.e., $\gamma(n) \leq \Gamma_\ell \cdot \gamma(\mathcal{N}^T)$.*

Note that the decision of advertisement is based on the capacity of $\mathcal{N}^T$, which may not be publicly known. However, we are assuming a node in $\mathcal{N}^T$ can estimate the size of the set. Nonetheless, a carefully chosen Γ_ℓ value would lead any rational node to advertise each transaction with overwhelming probability.

Corollary 2 is a result of the Lemma 3 where the greater than or equal to condition given in (9) is fixed to the equality. The reason for that is maximizing the reward of the round leader, which can be deducted from Theorem 1. Maximizing the round leader's reward is not necessary for the sake of the protocol, yet it would encourage the round leader to fulfill the block capacity. Otherwise, if intermediary nodes would earn more rewards, then miners may not have enough incentives to became a round leader.

4.3 Rewarding Function

We combine our findings and present the ideal rewarding function $\mathcal{RF}_{ideal}$ in the following theorem.

Theorem 1 *The following $\mathcal{RF}_{ideal}$ satisfies the desired properties of an ideal rewarding function for the transaction advertisement.*

$$\mathcal{RF}_{ideal} : \{F, \mathbf{P}\} \longrightarrow \{r_\ell^{[i]}\}_{i=1}^{\ell} : \sum_{i=1}^{\ell} r_\ell^{[i]} + r_\ell^{[NRL]} = F \text{ where}$$

$$r_\ell^{[i]} : \begin{cases} F \cdot \sum_{j=i}^{\ell} \left((-1)^{j-i} \cdot \binom{\ell-i}{j-i} \cdot \prod_{k=1}^{j-1}(1 - \Gamma_k)\right) & \text{for } 1 \leq i < \ell, \\ F \cdot \prod_{k=1}^{\ell-1}(1 - \Gamma_k) & \text{for } i = \ell. \end{cases}$$

Proof According to the Definition 1, an ideal rewarding function would satisfy incentive-compatibility and Sybil-proofness properties.

We formulate the Sybil-proofness condition in Eq. (5) and the incentive-compatibility condition in Corollary (2). By using (5) with $i = \ell$ and the corollary, we can compute the reward of a (potential) round leader with distance ℓ to the client:

$$r_\ell^{[\ell]} = r_{\ell+1}^{[\ell]} + r_{\ell+1}^{[\ell+1]} = \Gamma_\ell \cdot r_\ell^{[\ell]} + r_{\ell+1}^{[\ell+1]} \implies r_\ell^{[\ell]}(1 - \Gamma_\ell) = r_{\ell+1}^{[\ell+1]}$$
$$\implies r_\ell^{[\ell]} = r_{\ell-1}^{[\ell-1]}(1 - \Gamma_{\ell-1}) = \cdots = r_1^{[1]} \cdot \prod_{k=1}^{\ell-1}(1 - \Gamma_k) = F \cdot \prod_{k=1}^{\ell-1}(1 - \Gamma_k). \tag{10}$$

By combining Eqs. (10) and (6), we can also compute the reward of an intermediary node:

$$r_\ell^{[i]} = \sum_{j=i}^{\ell}(-1)^{j-i} \cdot \binom{\ell - i}{j - i} \cdot r_j^{[j]} = F \cdot \sum_{j=i}^{\ell}\left((-1)^{j-i} \cdot \binom{\ell - i}{j - i} \cdot \prod_{k=1}^{j-1}(1 - \Gamma_k)\right).$$

The rest of the fee is given to the next round leader, i.e., $r_\ell^{[NRL]} = F - R_\ell = F - \sum_{i=1}^{\ell} r_\ell^{[i]}$. □

4.4 *Tuning* Γ_ℓ

Γ_ℓ value is a crucial variable to assure that there will be an incentive to advertise a transaction until it reaches the whole blockchain network. As Γ_ℓ value increases, it will be easier to satisfy the incentive condition given in Eq. (9) since there will be more nodes having capacities less than $\Gamma_\ell \cdot \gamma(\mathcal{N}^T)$. On the other hand, the higher Γ_ℓ value, the lower reward remains for the rest of the advertisement path. From Eq. (10), it can be said that higher values of Γ_ℓ significantly reduces the reward of the round leader.

Assume that each new node connects to at least M nodes in the blockchain network. At the beginning of the advertisement, the client would advertise to its neighbors, thereby at least M different nodes would be aware of the transaction. For this case, $\Gamma_1 = 2/M$ would be a safe choice, which implies that at least half of the nodes satisfy the incentive condition in Eq. (9).

For the transition between Γ_ℓ and $\Gamma_{\ell-1}$, it is reasonable to assume that the number of nodes in $\mathcal{N}^T$ with distance ℓ to the client is probably more than those with distance $\ell - 1$. Therefore, in general, Γ_ℓ would be much less than $\Gamma_{\ell-1}$. For the first values of ℓ, an approximation in a random graph model would be $\Gamma_\ell \approx (\Gamma_1)^\ell$, $\Gamma_1 \in (0, 1)$.

4.5 On Countermeasures Against Forking Attacks

In the existing blockchain mechanisms, forking or censoring attacks can occur where the attacker does not accept a block including a high transaction fee and tries to rebuild the block with the same transactions to earn the fee [3]. The reason for the attack is that only the round leader who adds the transaction benefits from the transaction fee. Note that the previous proposals on the advertisement rewarding mechanisms do not address this issue.

In order to mitigate these forking attacks, transaction fees can be shared with the round leader of the next round [7]. In this way, nodes would also benefit from extending a block with a high transaction fee. Because of the difference in our model given in Definition 1 with the previous works, it is possible to reward the next round leader. Therefore, our rewarding function presented in Theorem 1 is compatible with the countermeasure against the forking attacks.

5 Conclusion

In this work, we investigated rewarding functions which are necessary to incentivize the rational nodes to participate in the transaction advertisement process. We formulate the properties of an ideal rewarding function, which are Sybil-proofness and incentive compatibility.

We introduced a new rewarding model where the total reward of the nodes in an advertisement path may vary with respect to the length of the path. With the help of the new model, we overcame the impossibility result given in [6] for Sybil-proofness in poorly connected networks. To the best of our knowledge, we present the first rewarding function which is compatible with any blockchain network model.

Our rewarding function divides the transaction fee among the nodes who advertise it, the current round leader and the next round leader. Since the next round leader also benefits from the transaction fee, the leader would not attempt to fork the chain to steal the transaction fees placed in the previous block. Thereby, unlike previous proposals, our rewarding function provides resistance against the forking attacks.

Open research questions. There are some open questions left which are beyond the scope of this work. The proposed rewarding mechanism requires well chosen Γ_ℓ values for the advertisement. In Sect. 4.4, we briefly discussed the trade-off between higher and lower values of Γ_ℓ. We left as a future work on deciding the optimum Γ_ℓ value which also depends on the blockchain network. In addition, our model does not take into account physical restrictions like the storage limit of the block size or the cost of the propagation of a transaction. An extension of the analysis regarding these real-world constraints would give a better approximation on the need of the rewarding function.

References

1. Abraham, I., Malkhi, D., Nayak, K., Ren, L., Spiegelman, A.: Solidus: an incentive-compatible cryptocurrency based on permissionless byzantine consensus. CoRR (2016). ArXiv:1612.02916
2. Babaioff, M., Dobzinski, S., Oren, S., Zohar, A.: On bitcoin and red balloons. In: Faltings et al. [8], pp. 56–73. https://doi.org/10.1145/2229012.2229022
3. Carlsten, M., Kalodner, H., Weinberg, S.M., Narayanan, A.: On the instability of bitcoin without the block reward. In: Proceedings of the 2016 ACM SIGSAC Conference on Computer and Communications Security, pp. 154–167. CCS '16. ACM, New York, NY, USA (2016). https://doi.org/10.1145/2976749.2978408
4. Drucker, F., Fleischer, L.: Simpler sybil-proof mechanisms for multi-level marketing. In: Faltings et al. [8], pp. 441–458. https://doi.org/10.1145/2229012.2229046
5. Emek, Y., Karidi, R., Tennenholtz, M., Zohar, A.: Mechanisms for multi-level marketing. In: Shoham, Y., Chen, Y., Roughgarden, T. (eds.) Proceedings 12th ACM Conference on Electronic Commerce (EC-2011), San Jose, CA, USA, June 5–9, 2011, pp. 209–218. ACM (2011). https://doi.org/10.1145/1993574.1993606
6. Ersoy, O., Ren, Z., Erkin, Z., Lagendijk, R.L.: Transaction propagation on permissionless blockchains: incentive and routing mechanisms. In: Crypto Valley Conference on Blockchain Technology, CVCBT 2018, Zug, Switzerland, June 20–22, 2018, pp. 20–30. IEEE (2018). https://doi.org/10.1109/CVCBT.2018.00008
7. Eyal, I., Gencer, A.E., Sirer, E.G., van Renesse, R.: Bitcoin-NG: a scalable blockchain protocol. In: Argyraki, K.J., Isaacs, R. (eds.) 13th USENIX Symposium on Networked Systems Design and Implementation, NSDI 2016, Santa Clara, CA, USA, March 16–18, 2016, pp. 45–59. USENIX Association (2016), https://www.usenix.org/conference/nsdi16/technical-sessions/presentation/eyal
8. Faltings, B., Leyton-Brown, K., Ipeirotis, P. (eds.): ACM Conference on Electronic Commerce, EC '12, Valencia, Spain, June 4–8, 2012. ACM (2012). http://dl.acm.org/citation.cfm?id=2229012
9. Gencer, A.E., Basu, S., Eyal, I., van Renesse, R., Sirer, E.G.: Decentralization in bitcoin and ethereum networks. CoRR (2018). http://arxiv.org/abs/1801.03998
10. Kaskaloglu, K.: Near zero bitcoin transaction fees cannot last forever. In: The International Conference on Digital Security and Forensics (DigitalSec2014), pp. 91–99. The Society of Digital Information and Wireless Communication (2014)
11. Kleinberg, J.M., Raghavan, P.: Query incentive networks. In: 46th Annual IEEE Symposium on Foundations of Computer Science (FOCS 2005), 23–25 October 2005, Pittsburgh, PA, USA, Proceedings, pp. 132–141. IEEE Computer Society (2005). https://doi.org/10.1109/SFCS.2005.63
12. Kroll, J.A., Davey, I.C., Felten, E.W.: The economics of bitcoin mining, or bitcoin in the presence of adversaries. In: Proceedings of WEIS, vol. 2013 (2013)
13. Li, C., Yu, B., Sycara, K.P.: An incentive mechanism for message relaying in unstructured peer-to-peer systems. Electron. Commer. Res. Appl. **8**(6), 315–326 (2009). https://doi.org/10.1016/j.elerap.2009.04.007
14. Nakamoto, S.: Bitcoin: a peer-to-peer electronic cash system (2008)
15. Pass, R., Shi, E.: Fruitchains: a fair blockchain. In: Schiller, E.M., Schwarzmann, A.A. (eds.) Proceedings of the ACM Symposium on Principles of Distributed Computing, PODC 2017, Washington, DC, USA, July 25–27, 2017, pp. 315–324. ACM (2017). https://doi.org/10.1145/3087801.3087809
16. Peters, G.W., Panayi, E.: Understanding modern banking ledgers through blockchain technologies: future of transaction processing and smart contracts on the internet of money. In: Banking Beyond Banks and Money, pp. 239–278. Springer (2016)
17. Sirer, E.G.: Bitcoin runs on altruism (2015). http://hackingdistributed.com/2015/12/22/bitcoin-runs-on-altruism/

18. Sompolinsky, Y., Zohar, A.: Bitcoin's underlying incentives. Commun. ACM **61**(3), 46–53 (2018). https://doi.org/10.1145/3152481
19. Wood, G.: Ethereum: a secure decentralised generalised transaction ledger. Ethereum Proj. Yellow Pap. **151** (2014)

PoolSim: A Discrete-Event Mining Pool Simulation Framework

Sam M. Werner and Daniel Perez

Abstract In Proof-of-Work cryptocurrencies, fair reward distribution within mining pools has become a popular area of research. Aside from a theoretical grounding, mining pool reward scheme research has commonly involved discrete event simulations of deterministic miner behaviour under different reward schemes. However, until now researchers have been left with the tedious task of developing their own mining pool simulation software, a rather time-consuming and potentially extensive undertaking, as miner behaviour becomes more complex. We present *PoolSim*, an open-source and very extensible discrete event simulation framework for modelling different behaviours of miners under any reward distribution scheme. By utilising this framework for different hypothetical mining scenarios, we showcase that mining pool reward scheme analysis indeed remains an exciting area for future research. Further, we believe that *PoolSim* will vastly increase productivity of researchers focusing on mining pools.

Keywords Mining pools · Discrete event simulation · Mining pool reward scheme analysis · Simulation tools

1 Introduction

Ever since Satoshi Nakamoto first introduced the concept of Bitcoin [9] in 2008, the cryptocurrency landscape has turned into a \$134 billion[1] digital gold mine. This can be accredited to certain features Bitcoin first introduced, perhaps the most notable, a

[1]Total cryptocurrency market capitalisation. Source: https://coinmarketcap.com. Accessed: 2019-03-06.

S. M. Werner (✉) · D. Perez
Imperial College London, London, UK
e-mail: sam.werner16@imperial.ac.uk

D. Perez
e-mail: daniel.perez@imperial.ac.uk

P. Pardalos et al. (eds.), *Mathematical Research for Blockchain Economy*,
Springer Proceedings in Business and Economics,
https://doi.org/10.1007/978-3-030-37110-4_12

decentralised consensus mechanism. The consensus mechanism is rooted in so-called *miners* trying to solve a computationally intensive cryptographic puzzle, referred to as the *Proof-of-Work* (PoW), where the difficulty is determined by the network. Miners are nodes which collect valid transactions while simultaneously trying to find a valid solution to the PoW problem. Solutions to the PoW are generated using modified versions of the block header to perform a preimage attack on the SHA-256 hash function in order to find a value which lies below some target threshold. The number of solution candidates, or hashes, a miner can compute per second denotes a miner's *hash rate*, typically expressed in megahashes[2] (MH/s) or gigahashes[3] per second (GH/s). The miner who finds a solution to the PoW is compensated for his invested computational effort in the form of newly minted units of the underlying cryptocurrency.

Mining remains a fundamental structural component for the effective workings of the decentralised consensus mechanism in Proof-of-Work cryptocurrencies, such as Bitcoin and Ethereum [1]. Although some other consensus mechanisms are being explored [12], PoW remains the most common consensus algorithm for cryptocurrencies, being used by 4 of the top 5 cryptocurrencies with the highest market capitalization [2]. However, rising difficulty levels of the cryptographic puzzles underpinning the mining process have posed severe constraints on the frequency of rewards paid to individuals trying to find PoW solutions. Hence, miners frequently pull their computational resources together to form *mining pools*, with the intention of reducing the high payout variance individual miners face. However, mining pool operators are faced with the non-trivial task of distributing rewards between pool participants in a fair manner, and in the past there has been much research [4–6, 8, 10, 13, 14] on the effectiveness of different mining pool reward schemes.

With the introduction of *PoolSim*, an open-source[4] discrete event simulation framework, researchers on mining pools and reward distribution schemes are no longer required to develop their own simulation software for modeling the behaviour of a miner under a particular mining pool reward scheme. To the best of our knowledge, there does not yet exist an actual simulation framework allowing users to fit each component to their needs by offering a high degree of extensibility, reliability, efficiency and good documentation. *PoolSim* is aimed at individuals interested in modeling different aspects of Proof-of-Work cryptocurrency mining pools with a focus on the reward distribution scheme employed.

The remainder of this paper is structured as follows. In Sect. 2 we explain cryptocurrency mining and outline the most common mining pool reward schemes. We subsequently examine some of the vulnerabilities identified in these reward schemes in Sect. 3. We explain the design of *PoolSim* in Sect. 4, prior to demonstrating its functionality by reproducing relevant existing academic work, as well as proposing potential avenues of future research in Sect. 6. We conclude in Sect. 7.

---

[2] 1 MH/s = 10^6 hashes per second.

[3] 1 GH/s = 10^9 hashes per second.

[4] https://github.com/samwerner/PoolSim. Accessed: 2019-03-02.

2 Preliminaries

In this section, we provide an overview of some of the most popular mining pool reward schemes, as examined in detail in [10, 14].

2.1 Solo Mining

Miners mining independently are generally referred to as *solo miners*. Finding a block with some constant hash rate h follows a Poisson distribution with the rate parameter $\lambda = \frac{h}{D}$, where D is the network difficulty. Hence, a solo miner with a hash rate h who mines for time period t has an expected revenue of

$$E[R] = \frac{htB}{D} \tag{1}$$

where B is the block reward. A financial risk remains for a solo miner, rooted in the high payout variance the miner is faced with.

2.2 Mining Pools

Mining pools are typically run by a centralised pool operator who issues so-called *shares*, or PoW problems with a difficulty d, which is lower than the network difficulty. Each share has a probability of $\frac{d}{D}$ of being a PoW solution. Hence, the number of expected shares per block is equal to

$$E[S] = \frac{D}{d}. \tag{2}$$

The pool operator's task is to check whether a submitted share is a solution to the PoW puzzle and therefore whether a block has been mined. In order to compensate for his efforts, the pool operator retains a fixed proportion f of each block reward B mined by the pool. The remainder of the block reward $(1 - f)B$ is distributed among the pool participants according to some reward distribution scheme implemented by the pool operator.

2.3 Mining Pool Reward Schemes

In Bitcoin, a block reward amounts to 12.5 BTC per block, whereas in Ethereum the block reward equals 2 ETH. In addition to receiving the transaction fees included in the block, in Ethereum, miners also receive rewards for mining or referencing uncle blocks. These are valid blocks that do not become the head of the longest chain. Ethereum rewards uncle blocks in order to incentivise miners to converge to a single chain, opposed to continue to mine on different branches in the event of a fork. The precise amount of the uncle reward is derived from the number of generations the uncle block lies away from the block which it is referenced by.

Fair distribution of block rewards within mining pools is not trivial. As multiple different reward distribution schemes have evolved over the years, we shall briefly discuss some of the most popular ones, as examined in more detail in [10, 14].

Proportional Payouts. The simplest and perhaps most intuitive mining pool reward scheme is the *proportional* reward allocation. In this scheme, block rewards are distributed proportionally between miners according to the number of shares each miner submitted for the current round. A *round* refers to the time period between two blocks being mined by a pool. Hence, if a miner submitted n shares during a round in which N shares have been submitted in total, the miner's payout would be equal to $\frac{n}{N}(1-f)B$.

Pay-Per-Share (PPS). In a *pay-per-share* scheme, the pool operator pays a miner $(1-f)pB$, where p is the probability that the share is a PoW solution, i.e. $\frac{d}{D}$. By doing so, the operator fully absorbs the miner's payout variance. Hence, an operator of a PPS pool could make profits on short rounds, while being exposed to high losses on long rounds. Typically, in order to compensate for this risk, PPS operators charge higher fees compared to other reward schemes. For understanding optimal reserve balances in a PPS pool we point the reader to Rosenfeld [10], who formally examines this.

Pay-Per-Last-N-Shares (PPLNS). Unlike many other traditional schemes, PPLNS abandons the concept of splitting rewards based on rounds, but rather distributes the block reward evenly among the last N shares submitted by miners, where N typically is a multiple of the network difficulty. This automatically takes away the incentive to only submit shares during the early period of a round. In a PPLNS pool, the expected reward normalised to a per round basis is found to be

$$E[R_i] = (1-f)B\frac{s_i}{N} \cdot \frac{N}{D} \tag{3}$$

where s_i is the number of shares submitted by miner i during the last N shares [10].

Queue-based (QB). The *queue-based*[5] reward scheme was introduced by Ethpool, a small Ethereum mining pool. In a queue-based mining pool, miners receive a number of *credits* equal to the difficulty of a share for each submitted share. When a block is

[5]Ethpool refers to this as a predictable solo mining pool, however, we shall employ the term "queue-based pool" as introduced in [14].

Table 1 The priority queue with miners of different sizes over a series of blocks

Position	Miner	Credits
1	Bob	140
2	Alice	130
3	Carol	70

(a) Before block i

Position	Miner	Credits
1	Alice	130
2	Carol	70
3	Bob	10

(b) After block i

Position	Miner	Credits
1	Alice	140
2	Carol	75
3	Bob	20

(c) After block i + 1

Position	Miner	Credits
1	Carol	75
2	Alice	65
3	Bob	20

(d) After block i + 1

mined by the pool, the full block reward[6] is allocated to the miner in the pool with the highest accumulated credit balance, namely the *top miner* in a priority queue. Subsequently, the top miner's credit balance is reset to the difference between his and the second highest credit balance in the pool. The reason for not resetting the credits of a top miner to zero is to provide an incentive for a miner to continue to perform work for the pool once he reaches the top position in the queue, opposed to switching to some other pool.

As first shown by [14], we provide a similar queue-based mining pool example in Table 1 showcasing that the credit resetting mechanism is non-uniform in the sense that the credits of top miners may be reset to differing balances. After a block has been mined by the pool (Table 1b), the top miner Bob has his credits reset to 10, the difference to Alice's credits. After each miner submits shares and receives credits, right before the next block is being mined (Table 1c), Alice is top of the queue. However, once the block has been mined (Table 1d), Alice is reset to a starting balance of 65 credits, a notably higher amount than Bob received. This suggests that Bob would indeed have been better off had he allowed Alice to bypass him in the queue in order to receive a higher starting balance. We shall examine potential vulnerabilities rooted in this non-uniform credit reset mechanism in the next section.

3 Reward Scheme Vulnerabilities

Several mining pool reward scheme vulnerabilities have been identified in the past. In this section we outline what the main reward scheme-targeted attacks are and how these differ from one another.

[6]Minus the pool operator fee.

3.1 *Block Withholding*

Research on possible attack scenarios between different mining pools has examined the effects of possible withholding attacks, whereby a mining pool mines in different mining pool and withholds blocks in order to cause direct harm to the pool [5, 6]. Furthermore, Schrijvers et al. [11] have shown that under proportional reward schemes, miners may deliberately hold on to a PoW solution for a temporary period of time in order to increase their payout, consequently harming the pool in which they mine.

3.2 *Pool-Hopping*

Apart from formally examining traditional reward schemes, Rosenfeld [10] also studies the effects of *pool-hopping*, whereby miners strategically decide in which pool to allocate their computational power in order to receive a reward higher than their fair share. Rosenfeld shows how unlike PPLNS, a proportional reward scheme is susceptible to pool-hopping attacks, as a miner is incentivised to leave the pool if a given round becomes too long due to the pool being unlucky. An examination of optimal pool-hopping behaviour and potential problems rooted in the prevention of such pool-hopping attacks has been conducted by [3, 4, 7].

3.3 *Queue-Based Manipulation Attacks*

Zamyatin et al. [14] examine potential vulnerabilities rooted in a queue-based mining pool and examine different strategies miners by the effects of different reward-increasing strategies a miner may pursue. These strategies are aimed at exploiting the non-uniform credit reset mechanism of the pool. As shown, in Sect. 2.3, a miner close to the top of the queue could be better off by allowing some other miner to surpass him in terms of total credits. By manipulating the queue order, a miner could aim to receive a higher credit starting balance once his credits have been reset. The authors identify three different actions a miner may take upon the submission of his share to manipulate the order of the queue, these being: share withholding, share donation and mining on a second wallet. For readability, we shall refer to the miner employing such behaviour as the *attacker*.

Share Withholding. The first action a miner may take to alter the queue order is to withhold computed shares from the pool operator. This has the effect that the attacker may be overtaken by some other miner with a higher credit balance in the queue. However, this approach suffers from inefficiency as the attacker performs work which is essentially lost. Furthermore, if the attacker is the top of the queue, there is no guarantee that the attacker will indeed be overtaken by some other miner in time before the round ends and his credits are reset.

Share Donation. In order to increase the likelihood of being bypassed by some miner, the attacker may donate his share to the address of the miner he intends to be overtaken by. The pool operator only receives the share and the address which should be credited for it and is therefore unable to make out donated shares. However, under this approach, the attacker suffers from missing out on performed work by giving shares away to other miners.

Second Wallet. In order to not miss out on performed work, an attacker may donate his shares to some other address, or wallet, of a set of addresses, which he controls in the same pool. Donating shares to a second address ensures that the attacker does not loose out on any performed work, while also being able to wait until he is overtaken by some other miner in the queue.

3.4 Queue-Based Uncle Mining

A recent analysis of queue-based reward schemes has been conducted by Werner et al. [13], who use mining pool simulations to reconstruct real-world observations of a miner exploiting the uncle block reward distribution mechanism of a small Ethereum mining pool. Under the examined scheme, an uncle reward was allocated to some miner on a random basis. However, this supposedly random allocation did not take into account miners' hash rates and was thus susceptible to Sybil attacks, where miners would spread their hash rate across a large number of so-called *uncle traps*, low-hash rate miners which exist for the sole purpose of receiving an uncle reward. The authors show how miners in this particular queue-based pool are incentivised to deliberately withhold valid shares from the pool operator until a block has been mined by the network, in order to produce an uncle block and harvest the associated uncle reward.

4 PoolSim: Design and Implementation

Our system is implemented in C++ and is designed to be highly configurable and extensible. The proposed system is primarily composed of a shared library `libpoolsim` which provides the core functionality of the simulator and an executable `poolsim` which takes a configuration file as input, allowing for easily running simulations. In this section, we describe the overall design of the system and provide some of the most relevant implementation details. We give a high-level overview of the design of the library in Fig. 1.

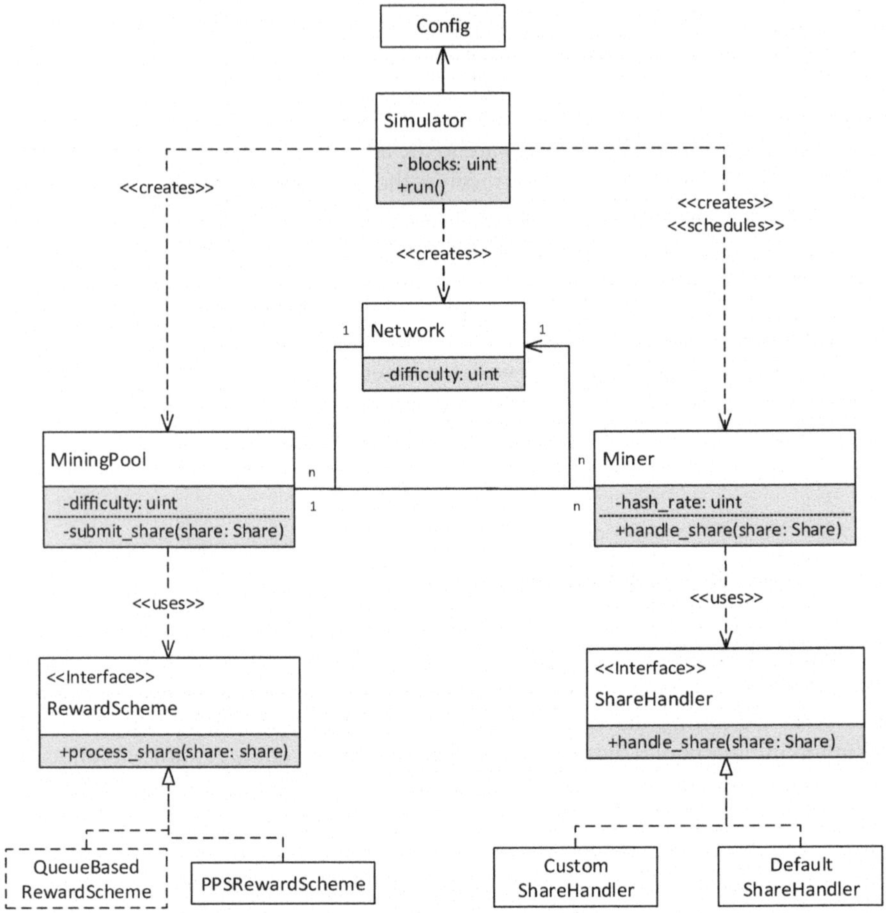

Fig. 1 Design overview of `libpoolsim`

PoolSim is a discrete-event simulator using a priority queue to store and execute scheduled share events. Share events represent shares generated by a miner, where the time intervals by which shares are submitted are constructed as a randomly generated exponentially distributed number with a rate parameter equal to $\lambda = \frac{h}{d}$, where h is the hash rate of a miner and d the share difficulty.

When a share is found, the miner is able to handle the share by using the provided share handler which can be configured or extended. The handler will usually submit the share to the pool it belongs to, but can choose any other behavior. Upon receiving a share, the mining pool delegates the share to the configurable reward scheme, which implements the logic for distributing rewards to miners. Once all this is done, the next scheduled miner processes the share it found, and this continues until the number of blocks in the simulation is reached. Each reward scheme and miner behavior can define its own metrics which are serialized with the final results.

Listing 1 PoolSim configuration

```
{
  "blocks": 10000,
  "seed": 120,
  "network_difficulty": 1000,
  "pools": [{
    "uncle_block_prob": 0.01,
    "difficulty": 10,
    "reward_scheme": {"type": "qb",
                      "params": {"pool_fee": 0.05}},
    "miners": {
      "generator": "random",
      "params": {
        "behavior": {"name": "default"},
        "hashrate": {"distribution": "normal",
                     "params": {"mean": 20, "variance": 5}},
        "stop_condition": {"type": "total_hashrate",
                           "params": {"value": 100}}
      }
    }
  }]
}
```

Configuration. Many different settings of PoolSim can be configured using a simple JSON configuration file. Rather than describing all the different parameters, we show a minimal sample configuration file in Listing 1.

This configuration runs a simulation for 10 000 blocks with a network difficulty of 1 000. The simulation contains a single queue-based mining pool with a share difficulty of 10 and a probability of its shares becoming uncle blocks of 1%. All the miners in the mining pool have the default behavior, which is to submit a share to the pool when found. In this example, the miners' hash rates are taken from a configurable normal distribution of mean 20 and variance 5, truncated at 0, as the hash rate cannot be negative. New miners are added to the pool until the total hash rate of the pool reaches 100.

Listing 2 Custom handler which withholds the share with a given probability

```cpp
// maybe_withhold_handler.h
#include <nlohmann/json.hpp>
#include <poolsim/share_handler.h>

class MaybeWithhold : public BaseShareHandler<MaybeWithhold> {
public:
  explicit MaybeWithhold(const nlohmann::json& args);
  void handle_share(const Share& share) override;
private:
  float withhold_prob;
};

// maybe_withhold_handler.cpp
#include "maybe_withhold_handler.h"

MaybeWithhold::MaybeWithhold(const nlohmann::json& args)
  : withhold_prob(args["withhold_prob"]) {}
```

```
void MaybeWithhold::handle_share(const Share& share) {
  if (std::rand() >= withhold_prob) {
    get_pool()->submit_share(get_address(), share);
  }
}

REGISTER(ShareHandler, MaybeWithhold, "maybe_withold")
```

Extending PoolSim. One of the most important features of PoolSim is its extensibility. There are two main parts which are designed to be very easily extended: the mining pool reward scheme and a miner's share handler. In both cases, a base class is provided and custom behavior can be implemented by subclassing it. In Listing 2, we demonstrate the simplicity of creating new behaviors by presenting a share handler which withholds the share with the probability given in the configuration file.

Once the desired behaviors are created, a new poolsim executable including these can be created by linking against `libpoolsim`. This also allows to integrate *PoolSim* to an existing C++ code base.

5 Simulations

In this section, we first use *PoolSim* to reproduce existing research that has used discrete-event simulations to examine profitability of different mining pool strategies. Additionally, we illustrate innovative use-cases of *PoolSim*, which to the best of our knowledge have not been examined before.

Simulation Setup. For the performed simulations, we assumed static network and share difficulties of 1 000 000 and 10 000, respectively. It should be noted that *PoolSim* allows users to define the logic for having dynamically self-adjusting network and/or share difficulties. The duration of each simulation was set to 100 000 blocks and a pool size of 1 000 miners has been assumed, unless stated otherwise. An uncle rate of 0% and a pool operator fee of 0%[7] have been assumed. For the attack scenarios in a queue-based pool we have set the condition that if the miner in the position after the attacker in the queue, has a total credit balance equal to 90% or more of the attacker's credits, the attacker executes some deterministic strategy. We define the attacker to be a miner with a total hashrate of 15 GH/s. All simulations are run on an Intel i7-8550U CPU clocked at 1.80 GHz, with 16 GB of RAM clocked at 1066 MHz. In the current state of our implementation, the simulator is single threaded and we therefore only use one of the 8 threads available on the CPU.

[7] As the performed simulations did not involve a PPS reward scheme under which an operator fee would indeed be relevant, we decided to omit this variable.

Table 2 Blocks mined and rewarded under solo mining compared to mining in a PPLNS and QB two-miner scenario

Miner	Block Ratio			Avg. performed work per block		
	Solo	PPLNS	QB	Solo	PPLNS	QB
Large	1.0	1.000	0.993	1 000 974	1 000 827	1 008 227
Small	1.0	0.999	1.072	1 000 087	1 001 555	932 991

5.1 Existing Research

We use *PoolSim* in an attempt to reproduce some of the key contributions of Zamyatin et al. [14].[8] In order to compare a miner's payouts of mining in a PPLNS scheme to mining in a queue-based scheme, [14] construct and evaluate the performance of a two-miner case, in which a small 1 GH/s miner and a large 10 GH/s miner mine under each scheme. Furthermore, the authors identify attack strategies (Sect. 3.3) specific to a queue-based pool, which may be employed by a large miner to potentially increase his payout in a two-miner scenario.

We reconstruct and simulate[9] the normal two-miner case, in which both miners mine honestly absent any attack strategies.

Two-miner Case: Normal. Zamyatin et al. [14] find that when comparing the performance between miners in the two-miner case under different reward schemes that the small miner is in fact performing less work per block than the large miner in a queue-based pool. The authors also show that the large miner would have been better off mining in a PPLNS pool (best option).

In Table 2, we show our results of the reconstructed two-miner case. These are inline with the findings of [14], as it can be seen that the large miner performs a notable amount of more work than the small miner in the queue-based pool. This is reflected by the ratio of blocks received to blocks mined for the large miner in the queue-based pool, which is worse than had he mined in a PPLNS pool, or solo. The relatively high block ratio of the small miner for the queue-based pool is also inline with the findings by [14], indicating how the small miner benefits from the large miner absorbing the variance during lucky and unlucky streaks of the pool.

Two-miner Case: Attack. When looking at the attack strategies of share withholding, tactical donation of mining power and use of second wallets, the findings by Zamyatin et al. show that the optimal strategy for a large miner is the tactical donation of mining power in a two miner scenario.

We present the results for the two-miner attack scenarios in Table 3. As first discovered by [14], we also find that the tactical donation of mining power is the

[8] We would like to remark that *PoolSim* has the capabilities of simulating each of the attack scenarios presented in Sect. 3, however, for brevity we shall only focus on the work by [14].

[9] It should be noted that we have made small modifications to the simulation setup compared to the original work for performance gains.

Table 3 Attack simulation results in a two-miner scenario

Attack strategy	Prop. of avg. credits lost	Miner	Avg. performed work per block	Blocks		
				Rewarded	Mined	Ratio
Share withholding	0.188	Attacker	970 208	73 133	71 164	1.028
		Victim	1 295 488	7005	8974	0.781
Tactical donation of mining power	0.188	Attacker	972 887	91 192	88 610	1.029
		Victim	1 290 845	8 808	11 390	0.773
Using a second wallet	0.246	Attacker	1 105 026	82 538	82 181	1.004
		Victim	926 686	9833	9 121	1.078

most successful strategy. By pursuing this strategy, the attacker is able to increase his block ratio to 1.029, compared to 0.993 in the normal scenario presented in Table 2, and thereby compensate for his initial loss.

It should be noted that our simulation configuration deviates from the work done by [14] when looking at the share withholding strategy. As a miner does not submit the share he withholds, the share is essentially lost, even if this share is a valid solution to the PoW. Therefore, the attacker will find less blocks during the same time period as for the other scenarios, in which no work is lost. In [14], the length of the share withholding is extended to equal the same duration as all the other scenarios. However, we believe that not submitting shares and thus mining fewer blocks poses the risk of being worse off compared to submitting all shares.

Our findings for the effectiveness of the second wallet strategy deviate slightly from the results presented in the original work, as the victim in our simulation performed better than the attacker. However, this difference may be explained by possibly differing implementations of the simulators used. Overall, we were able to reproduce the key findings of [14], where the most effective strategy is the tactical donation of mining power, followed by the withholding of shares and the use of a second wallet, respectively.

Simulation Performance. In Table 4, we present the execution time of the different simulations performed for the two-miner scenario. It is worth noting that using our

Table 4 Execution time of the different simulated two-miner scenarios

Simulation	Runtime (ms)
PPLNS	5 310
Queue based	4 468
Share withholding	6 528
Tactical donation of mining power	6 805
Using second wallet	7 721

Table 5 Multi-miner simulation scenarios

Scenario	Avg. credit lost	Avg. performed work per block	Blocks		
			Rewarded	Mined	Ratio
PPLNS	NA	1,004,706	488.31	488	1.001
QB	0.0020	993,137	494.00	488	1.012
QB with share donation	0.0020	999,205	491.00	485	1.012

Table 6 Execution time of the different simulated multi-miner scenarios

Simulation	Runtime (ms)
PPLNS	331 957
Queue based	158 292
Share donation	166 464

implementation, the ratio between the network difficulty and the pool difficulty influences greatly the speed of execution, as it changes the number of shares which must be generated before finding a block. As this ratio does not influence the behavior of the attacks we are checking for, we decide to keep it low in order to speed-up the simulations. In our simulations, this ratio is of 100, compared to about 20 000 for most real-world pools.

5.2 *Normal Multi-miner and Attack Scenarios*

In the previous subsection, we have examined how *PoolSim* can be used to reconstruct and examine the effectiveness of different queue-based attack scenarios. To receive further insights into the dynamics of queue-based mining pools we compare the performance of an attacker between mining in different pools containing 1 000 miners each. Table 5 shows the *PoolSim* execution time per simulated scenario (Table 6).

When comparing the payouts of a miner under a PPLNS scheme to a QB scheme, we find that the attacker did in fact receive a slightly higher number of blocks in the QB pool (494) than in the PPLNS pool (488.31). We note that the attacker received in both pools a number of blocks higher than the number of blocks he actually mined. These types of analyses have indeed also already been conducted by [14]. However, we additionally simulate a scenario in which the attacker pursues the tactical donation of mining power strategy in a multi-miner pool. We selected the aforementioned strategy as this was the most effective one in the two-miner case. We find that the attacker received less blocks (491) than had he mined honestly in the QB pool. The reason as to why the share donation strategy did not have any noticeable effect in this multi-miner pool is rooted in the existing credit differences between miners. The tactical donation of shares strategy requires certain credit differences to exist, as otherwise giving away shares does not pay off over time. In order to measure the extent of such differences in a given pool, we turn to the average proportion of credits lost per round. For a given round, this measure is expressed as the number of credits of the second miner as a proportion of the total sum of credits of the pool. Looking at this proportion, we find that in the scenario of share donation in the two-miner case, the average proportion of credits lost amounts to 0.188. Interestingly, for the scenario of share donation in a multi-miner case, we find that this figure amounts to 0.002, and is thus significantly lower. In fact, we suggest that the average credit loss is one of the main variables which can be taken advantage of by an attacker. For example, in a simulation with no attacker, the average credit lost is 0.245. Table 3 shows that successful attacks manage to reduce this average.

5.3 Queue-Based Pool-Hopping

This simulation demonstrates *PoolSim*'s capability to simulate condition-based pool-hopping scenarios. As briefly discussed in Sect. 3, several analyses on the effects of pool-hopping, as well as on which reward schemes are vulnerable to it, exist. However, we note that there have been no studies on the feasibility of pool-hopping between queue-based mining pools. Hence, we construct and examine a *proof-of-concept* pool-hopping scenario, where a miner submits shares to a pool on the basis of the luck of the pool. Pool luck for a given round r can be defined as

$$l_r = \frac{S_E}{S_A} \cdot 100 \tag{4}$$

where S_E is the number of expected shares per round and S_A is the number of actual shares submitted per round.

We construct two mining pools and add a conditional hopping, stating that the attacker will leave the current pool he is in if the number of shares submitted by the pool for the current round is twice the amount as expected, or $l = 50\%$.

We find that the attacker received 251 blocks in total from hopping between both pools, while receiving 249 blocks when mining only in one pool. However, this is based on a rather simple set up, as both pools have log normal hash rate distributions, no attackers and are rather identical. Nonetheless, we have successfully shown that *PoolSim* can be used for simulating simple, as well as more complex conditions.

6 Future Research Using PoolSim

We believe that *PoolSim* can facilitate research on areas of mining pool reward schemes, such as fairness, vulnerabilities and attacks. As there has been the least amount of academic research on the queue-based reward scheme, we shift our focus to this scheme when pointing towards areas of future research. From the few simulations examined in the previous section, we were able to make some interesting observations. We showed that despite working in a two miner scenario, queue-based attack strategies are not necessarily nearly as effective in a multi-miner pool. This is presumably caused by the pool size and the hash rate distribution of the pool, as these variables directly affect the credit differences between miners in the pool.

Even though pool-hopping on a luck basis between two queue-based pools did not provide any novel insights, we were able to successfully demonstrate the powerful conditional pool-hopping functionality of *PoolSim*. A feature, which, to the best of our knowledge, has not been implemented and utilised elsewhere. Hence, with regards to employing *PoolSim* for future research, one could with very little effort construct a conditional pool-hopping scenario between multiple queue-based mining pools and add a more complex condition for submitting shares. An example of such a condition would be to find the optimal pool out of a set of queue-based mining pools in terms of highest average proportion of credits lost per round. Furthermore, this conditional logic could be extended to also employ a strategy such as the tactical donation of shares in order to actually exploit large credit differences when they occur. Such an exploratory approach could provide further insights into hash rate distributions of queue-based mining pools and their implications for the effectiveness of different attacks targeting the reset mechanism.

A additional area of future interest could lie in the game-theoretic aspects and analysis of multi-attacker scenarios in different pool constellations. Even though we only used the framework to look into single-attacker scenarios in two and multi-miner settings, *PoolSim* can be used to simulate multi-attacker scenarios. This could provide stimulating insights into examining the effects of mining scenarios in large pools, where multiple (or perhaps only) attackers exist, all pursuing the same or different attack strategies.

7 Conclusion

Examining the exploitability of potential vulnerabilities embedded within different existing rewards schemes employed by mining pools has evolved into an interesting area of research within PoW cryptocurrencies. In this paper, we have introduced *PoolSim*, a simulation framework targeted for academics or anyone with an interest in studying and examining incentive and security mechanisms of mining pool reward schemes. *PoolSim* allows for a high degree of customisation, where new reward schemes could easily be implemented and tested, or the effectiveness of new attack strategies can be assessed accurately. *PoolSim*, finally allows researchers not having to deal with unnecessarily time-consuming and complex implementation tasks. We have provided an overview of the design of *PoolSim*, while also having demonstrated the framework's functionality by reconstructing relevant academic research. Furthermore, we have used *PoolSim* to point out several potential areas of future work.

Acknowledgements The authors would like to thank Paul Pritz for helpful feedback. This research is supported by funding from the Brevan Howard Centre for Financial Analysis.

References

1. Buterin, V.: Ethereum: a next-generation smart contract and decentralized application platform. https://github.com/ethereum/wiki/wiki/White-Paper (2014). Accessed 22 Aug 2016
2. Cap, C.M.: Coin market cap. https://coinmarketcap.com (2019). Accessed 05 Apr 2019
3. Chávez, J., Rodrigues, C.: Hopping among pools in the bitcoin mining network. SIJ Trans. Comput. Netw. Commun. Eng. (CNCE) **3**(2), 22–27 (2015)
4. Chávez, J.J.G., da Silva Rodrigues, C.K.: Automatic hopping among pools and distributed applications in the bitcoin network. In: 2016 XXI Symposium on Signal Processing, Images and Artificial Vision (STSIVA), pp. 1–7. IEEE (2016)
5. Courtois, N.T., Bahack, L.: On subversive miner strategies and block withholding attack in bitcoin digital currency. arXiv:1402.1718 (2014)
6. Eyal, I.: The miner's dilemma. In: 2015 IEEE Symposium on Security and Privacy (SP), pp. 89–103. IEEE (2015)
7. Lewenberg, Y., Bachrach, Y., Sompolinsky, Y., Zohar, A., Rosenschein, J.S.: Bitcoin mining pools: a cooperative game theoretic analysis. In: Proceedings of the 2015 International Conference on Autonomous Agents and Multiagent Systems, pp. 919–927. International Foundation for Autonomous Agents and Multiagent Systems (2015)
8. Luu, L., Saha, R., Parameshwaran, I., Saxena, P., Hobor, A.: On power splitting games in distributed computation: the case of bitcoin pooled mining. In: Proceedings of the 28th IEEE Computer Security Foundations Symposium (CSF 2015), pp. 397–411. IEEE, IEEE Computer Society, Verona, Italy (2015)
9. Nakamoto, S.: Bitcoin: a peer-to-peer electronic cash system. Dec 2008, https://bitcoin.org/bitcoin.pdf. Accessed 01 Aug 2015
10. Rosenfeld, M.: Analysis of bitcoin pooled mining reward systems. arXiv:1112.4980 (2011). Accessed 16 Oct 2018
11. Schrijvers, O., Bonneau, J., Boneh, D., Roughgarden, T.: Incentive compatibility of bitcoin mining pool reward functions. In: Financial Cryptography and Data Security (2016)
12. Wang, W., Hoang, D.T., Xiong, Z., Niyato, D., Wang, P., Hu, P., Wen, Y.: A survey on consensus mechanisms and mining management in blockchain networks. arXiv:1805.02707 (2018)
13. Werner, S., Pritz, P., Zamyatin, A., Knottenbelt, W.: Uncle traps: harvesting rewards in a queue-based ethereum mining pool. Cryptology ePrint Archive preprint: 2019/070. https://eprint.iacr.org/2019/070.pdf (2019). Accessed 02 Mar 2019
14. Zamyatin, A., Wolter, K., Werner, S., Harrison, P.G., Mulligan, C.E., Knottenbelt, W.J.: Swimming with fishes and sharks: beneath the surface of queue-based ethereum mining pools. In: 2017 IEEE 25th International Symposium on Modeling, Analysis, and Simulation of Computer and Telecommunication Systems (MASCOTS), pp. 99–109. IEEE, IEEE Computing Society, Banff, Canada, Sept 2017. https://doi.org/10.1109/MASCOTS.2017.22

Oceanic Games: Centralization Risks and Incentives in Blockchain Mining

Nikos Leonardos, Stefanos Leonardos and Georgios Piliouras

Abstract To participate in the distributed consensus of permissionless blockchains, prospective nodes—or *miners*—provide proof of designated, costly resources. However, in contrast to the intended decentralization, current data on blockchain mining unveils increased concentration of these resources in a few major entities, typically *mining pools*. To study strategic considerations in this setting, we employ the concept of *Oceanic Games* [27]. Oceanic Games have been used to analyze decision making in corporate settings with small numbers of dominant players (shareholders) and large numbers of individually insignificant players, *the ocean*. Unlike standard equilibrium models, they focus on *measuring the value (or power) per entity and per unit of resource* in a given distribution of resources. These values are viewed as strategic components in coalition formations, mergers and resource acquisitions. Considering such issues relevant to blockchain governance and long-term sustainability, we adapt oceanic games to blockchain mining and illustrate the defined concepts via examples. The application of existing results reveals incentives for individual miners to merge in order to increase the value of their resources. This offers an alternative

Stefanos Leonardos and Georgios Piliouras acknowledge that this work was supported in part by the National Research Foundation (NRF), Prime Minister's Office, Singapore, under its National Cybersecurity R&D Programme (Award No. NRF2016NCR-NCR002-028) and administered by the National Cybersecurity R&D Directorate. Georgios Piliouras acknowledges SUTD grant SRG ESD 2015 097, MOE AcRF Tier 2 Grant 2016-T2-1-170 and NRF 2018 Fellowship NRF-NRFF2018-07.

N. Leonardos
National and Kapodistrian University of Athens, Panepistimioupolis, 161 22 Zografou, Greece
e-mail: nleon@di.uoa.gr

S. Leonardos (✉) · G. Piliouras
Singapore University of Technology and Design, 8 Somapah Rd, Singapore 487372, Singapore
e-mail: stefanos_leonardos@sutd.edu.sg

G. Piliouras
e-mail: georgios@sutd.edu.sg

P. Pardalos et al. (eds.), *Mathematical Research for Blockchain Economy*, Springer Proceedings in Business and Economics,
https://doi.org/10.1007/978-3-030-37110-4_13

perspective to the observed centralization and concentration of mining power. Beyond numerical simulations, we use the model to identify issues relevant to the design of future cryptocurrencies and formulate prospective research questions.

Keywords Blockchain · Cryptocurrencies · Resources · Mining pools · Oceanic games · Values

1 Introduction

Decentralization is a core element in the design of permissionless blockchains. To participate in the blockchain consensus mechanisms, prospective network nodes—also called *miners*—need to provide proof of some costly resource. This resource may be computational power in protocols with Proof of Work (PoW) selection mechanisms, [15, 28], or coins of the native cryptocurrency in Proof of Stake (PoS) selection mechanisms, [5, 8]. Under default conditions, the selection is proportional to miners' resources and hence, it depends on their actual distribution. An integral assumption in the security philosophy of permissionless blockchains is that the network of mining nodes remains "sufficiently" decentralized and distributed. In the extreme case, *sufficiently* means that no single entity holds 50% or more of the resources but in practice much more fragmentation may be desired to safeguard the safety properties of the underlying protocol [3, 11, 22].

With this in mind, the picture illustrated in Table 1 is disconcerting. Table 1 shows the distribution of blocks among miners in the two largest[1] cryptocurrencies, Bitcoin and Ethereum, and indicates that the desired assumption of a highly decentralized (and distributed) network is currently not satisfied. As can be seen, the vast majority of mining resources is concentrated in a small number of "major" nodes or *mining pools* in which individual miners join forces to reduce the variance of their payments [14, 34]. The rest is scattered among a large number of minor and individually insignificant miners. The discrepancy between the intended distribution and the concentration of resources that is observed in practice raises some questions. What is the actual power of such pools or major miners to influence the evolution of the blockchain? Does this distribution create incentives for mergers and formation of coalitions (cartels) that will seize control of the majority of resources and manipulate the blockchain [24, 26]? What strategic considerations arise and what are their implications on blockchain governance and long-term sustainability?

Similar questions have been examined by conventional economics in the context of corporate governance. To study interactions between shareholders with various degrees of power in particular, [27] developed the model of *Oceanic Games*. These are games featuring a mixture of few large players (shareholders) and a continuum of infinitesimal players, called the *ocean*, each of which holds an insignificant fraction of corporate shares. The resemblance with blockchain mining—with

[1] In terms of market capitalization, cf. coinmarketcap.com.

Table 1 Distribution of the blocks mined in the Bitcoin and Ethereum blockchains. Mining is dominated by few major miners, typically mining pools, numbered from 1 to 10 and a great number of minor players in the "Unkown/other" category. *Source* blockchain.com and etherscan.io, 5 March 2019

	Bitcoin		Ethereum	
	Entity (Pool)	Blocks (%)	Entity (Pool)	Blocks (%)
1.	BTC.com	18.2	Ethermine	28.2
2.	AntPool	14.7	Sparkpool	21.4
3.	F2Pool	12.6	Nanopool	12.6
4.	SlushPool	10.1	F2Pool_2	12.4
5.	BTC.TOP	7.9	MiningPoolHub_1	5.6
6.	ViaBTC	7.9	DwarfPool_1	1.9
7.	DPOOL	4.1	PandaMiner	1.8
8.	BitFury	2.3	firepool	1.6
9.	BitClub Network	2.3	Address_1	1.4
10.	Bitcoin.com	1	MinerallPool	1.1
	Unknown/other	**18.9**	**Unknown/other**	**12.0**

shares corresponding to units of mining resources—is apparent. Our goal in this paper is to explore incentives in blockchain mining from the perspective of Oceanic Games and complement existing studies that focus on safety and security related issues [9, 13, 29].

The central idea in the literature of Oceanic Games is the measurement of a *value* for each entity and for each *unit of resource* given the distribution of resources among shareholders. The concept of *value* is considered as a powerful tool in the theory of decision making [2, 31, 33] and [32]. For instance, if a miner holds 51% of the total resources, then each of her units is worth much more than if she holds only 49% of the total resources, since in the former case, the entirety of her shares gives her absolute control over the blockchain. Similar, but maybe less obvious considerations, arise also in intermediate cases. If a miner holds 49% of the resources and a second miner holds 2% of the resources, then both miner's resources value higher than in the case in which the first miner only holds 47% of the resources, since in the former case, the two miners may collude and jointly seize control of the blockchain.

Motivated by these considerations, we adapt the model of Oceanic Games from [27] on blockchain mining. Our aim is to measure the *value of mining resources per miner and per unit of resource* as a strategic component in the process of power gain and coalition formation between mining nodes. With this approach, we shift our attention from safety attacks and equilibration models, [12, 20], to the understanding of incentives related to the distribution and acquisition of protocol resources. The analysis of these issues is relevant to the broader subjects of long term sustainability and blockchain governance [7].

Based on the above, our contribution in the present paper can be summarized in the following points

- We model instances of blockchain mining as Oceanic Games: the discrete set of large players corresponds to the large mining nodes, typically mining pools, and the continuum of infinitesimal oceanic players to the remaining, individual miners, cf. Fig. 1. Conveniently, the resulting model does not depend on the underlying selection mechanism (PoW, PoS or similar) or consensus protocol and hence can be used for the study of resource acquisition, strategic interactions, coalition formations (mergers) and governance related issues in a broad spectrum of permissionless blockchains [4, 9, 10, 16, 18, 21, 30]. We extend an example of [27] to illustrate the defined concepts in blockchain context.
- The application of existing results uncovers incentives for the formation of mergers between miners. Starting from an initial distribution in which the oceanic players control the majority of resources, we use simulations to show that this holds in two instances: first, in the formation—crystallization—of a coalition out of the ocean and second, in the exogenous acquisition of additional resources by a group of individual miners who, nevertheless, have the ability to coordinate their actions (collude). In both cases, the value of the miners' resources is higher when they act as a single entity rather than individual, oceanic players. This result provides an alternative perspective to the observed centralization in cryptocurrency mining, cf. Table 1.
- Further numerical simulations demonstrate that the above conclusions do not hold in the whole range of parameters. Instead, the dynamics of coalition formations and entry barriers are shown to depend on the current distribution of mining resources among major miners and the ocean.
- Finally, we use this model to raise issues relevant to the design of future cryptocurrencies and formulate prospective research questions.

In general, the present paper can be seen as a first step towards the application of the Oceanic Game concept in blockchain mining. Beyond some first insight, the extent to which this model can provide further results in the issues of (de)centralization, blockchain governance and long-term sustainability is yet to be fully understood.

1.1 Outline

The rest of the paper is structured as follows. In Sect. 2 we present the model of Oceanic Games and give an example to illustrate the defined notions. Revelant results from [27] and their application in blockchain settings are shown in Sect. 3 along with numerical simulations. In Sect. 4, we raise related issues and research questions and discuss limitations of the current approach. Section 5 concludes the paper.

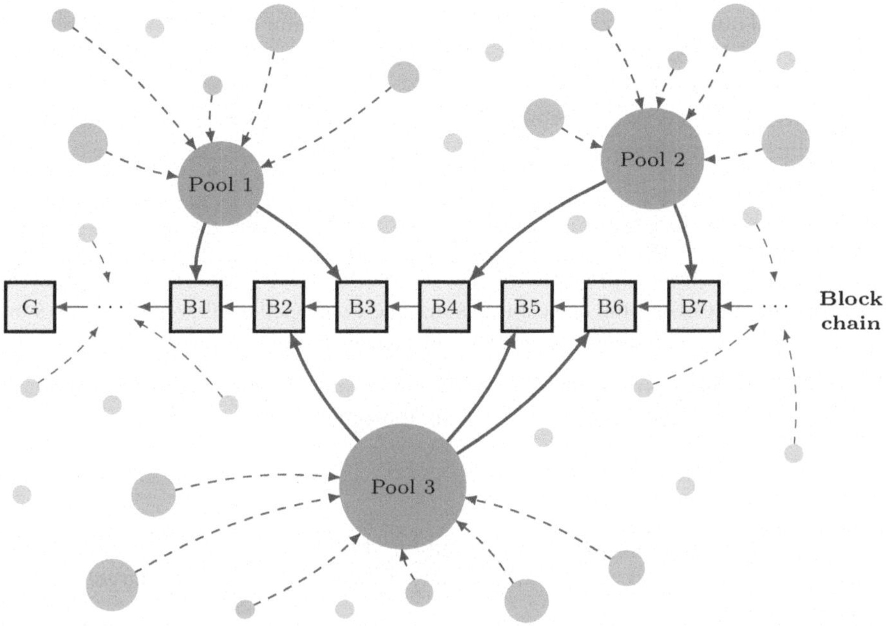

Fig. 1 Illustration of centralization in blockchain mining. Miners join forces in few major mining pools (blue), $M = \{1, 2, \ldots, m\}$, which dominate the mining process. The remaining small miners (light red)—or the ocean, I—mine individually

2 The Model: Oceanic Games on the Blockchain

The current model adjusts the notation and terminology of [27] in a standard blockchain setting.

Miners: The *miners* are the physical entities that participate in the block proposal and creation process. The term is used here in the broadest sense and depending on the underlying protocol and selection mechanism, it may refer to "conventional" miners as in PoW, [28], or to *virtual miners* as in PoS or other alternative forms [8]. The set of miners consists of two distinctive components

- A finite, discrete set $M = \{1, 2, \ldots, m\}$ of major miners or mining pools.
- An interval $I = [0, 1]$ of infinitesimal miners. We refer to I as the *ocean* and to miners in I, as *oceanic players*. We only consider subsets $U = [u_1, u_2] \subseteq I$ of the ocean I, e.g. $U = [0.1, 0.5]$, and not individual oceanic players.

Resources: To participate in the distributed consensus, each miner needs to provide proof of some designated, costly resource. This may be a physical or digital asset such as computational power in PoW or native coins in PoS mechanisms, respectively. To describe these resources, we use following notation

- A set of real numbers $r_1, r_2, \ldots, r_m \geq 0$, where r_i denotes the amount of resources of miner $i \in M$. For any subset $S \subseteq M$, we will write $r(S) = \sum_{i \in S} r_i$ to denote the total resources of miners in S.
- A positive constant $\alpha > 0$ which denotes the total resources of the ocean I. Accordingly, any subset $U = [u_1, u_2] \subseteq I$ controls $\alpha \cdot |U|$ of resources where $|U| = u_2 - u_1$.

  Based on the above, the total protocol resources R are equal to $R := \alpha + \sum_{i \in M} r_i$. While resources change over time, in the present analysis, we will focus on a single period or a static setting and hence, unless indicated otherwise, our notation is independent of the time t. Resources may be expressed as absolute numbers or percentages but this will be made explicitly clear from the context.

Blockhain Oceanic Games: Given the above, a *blockchain oceanic game* Γ is defined by a *majority quota*, $q \geq 0$, using the symbol[2]

$$\Gamma := [q; r_1, r_2, \ldots, r_m; \alpha]$$

with the following interpretation: a coalition of miners $C := S \cup U$ with $S \subseteq M$ and $U \subseteq I$ wins in the game Γ, if and only if its total resources are larger than or equal to q, i.e., if

$$r(C) := r(S) + \alpha \cdot |U| \geq q$$

Addition of resources: Given an oceanic game Γ, we want to study the situation in which new entities acquire resources and enter the protocol. For this, we will use the notation Γ^+ with

$$\Gamma^+ := \left[q; r_1, r_2, \ldots, r_m, r_{m+1}; \alpha\right]$$

and $M^+ = \{1, 2, \ldots, m, m+1\}$. In words, Γ^+ results from Γ by the addition of a new major player $m+1$ with exogenous resources $r_{m+1} > 0$. Similar notation can be used to denote the formation of a new entity *crystallizing* out of the ocean. In this case, $\Gamma^+ := \left[q; r_1, r_2, \ldots, r_m, r_{m+1}; \alpha - r_{m+1}\right]$ for some $r_{m+1} > 0$, and $M = \{1, 2, 3, \ldots, m+1\}$.

Values: The first core functionality of the present model is to calculate a *value* φ_i for each major miner $i \in M$ and one value Φ for the entirety of the oceanic players, also referred to as the *oceanic value*. Each miner's value depends on that miner's share of resources *and* on the total distribution of the remaining resources among the rest of major and oceanic miners. To define the miner's values $\varphi_i, i \in M$ and the oceanic value Φ, let $X_1, X_2, \ldots, X_m$ be independent random variables uniformly distributed on $I = [0, 1]$. For each $x \in I$, let $r(x) := \sum_{j \in M} r_j \cdot \mathbf{1}\{X_j < x\}$, where $\mathbf{1}\{X_j < x\} = 1$ if $X_j < x$ and 0 otherwise (indicator

---

[2]The notation is common in the literature of *weighted voting games*, see [25, 31, 32] for a more related application. Also, in most cases, we will be interested in $q = 0.5$ or 50% but the current model applies to any q of interest.

function). Then, the *value* of miner $i \in M$ is defined by

$$\varphi_i := \mathbb{P}\mathrm{rob}\left[r\left(X_i\right) + \alpha X_i < q \le r\left(X_i\right) + \alpha X_i + r_i\right] \tag{1}$$

and the oceanic value by $\Phi := 1 - \sum_{i \in M} \varphi_i$. Intuitively, the value φ_i is the probability that miner i will be the crucial entity to turn a random coalition of miners from losing (total resources of the coalition without i are less than q) to winning (total resources of the coalition with i are equal to or greater than q).[3]

Value-per-unit of resource: The second functionality of this model is to determine the *value per-unit of resource* or power ratio, υ_i, for each player $i \in M$, which is defined by

$$\upsilon_i := \varphi_i / r_i \tag{2}$$

Similarly, the *value per oceanic unit of resource* or oceanic power ratio, υ_{oc}, is equal to $\upsilon_{\text{oc}} := \Phi / \alpha$.

2.1 An Example: Why Values and Not Shares?

We illustrate the above with the help of an example adapted from [27, Section 6]. We consider a mining situation with two major mining entities or pools, $M = \{1, 2\}$, and the simple majority quota $q = 0.5$ represented by the following game $\Gamma = [0.5; r_1, r_2; \alpha]$, where $\alpha = 1 - r_1 - r_2$. The 0.5 or 50% quota corresponds to cntrol of the majority of protocol resources and hence, of the blockchain as a whole. In this game a coalition S wins, if $r(S) \ge 0.5$, i.e., if it occupies 50% or more of the protocol resources.[4]

All possible resource configurations (r_1, r_2, α) are illustrated in Fig. 2. The horizontal and vertical axes represent miner 1's and miner 2's fraction of the resources, respectively. Their possible combinations are divided in 4 inner regions, $\Delta_i, i = 1, 2, 3, 4$. Region Δ_1 contains all configurations for which the combined resources of both major miner are less than 50%, i.e., $r_1 + r_2 \le 0.5$. In this case, the majority of resources is controlled by oceanic players. However, the ocean is not actually "in control", since, by assumption, there is no coherence nor organizational structure between oceanic players. The explanation of regions Δ_2, Δ_3 and Δ_4 is similar and is briefly given in the legend of Fig. 2.

Using (1), the value φ_1 of the first major miner is given by

[3] For more details and the probabilistic derivation of these values, we refer to [27].

[4] Due to continuity properties, there is no difference between using the q = 50% quota or symbolically, the q = 51% quota, as is common in the related literature [11, 22, 28].

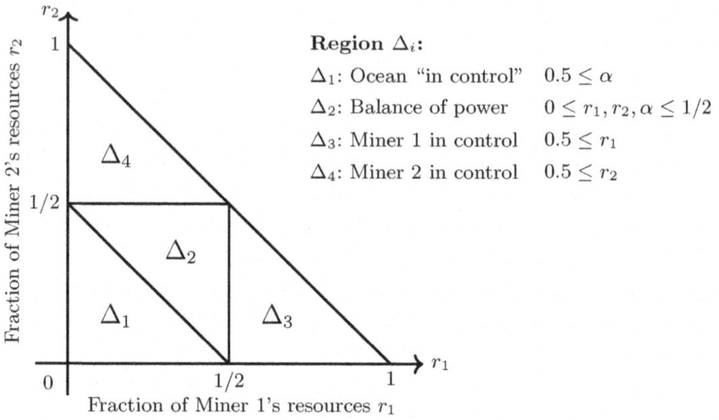

Fig. 2 All possible configurations in the distribution of resources (r_1, r_2) between 2 major miners and the ocean, α

$$\varphi_1 = \begin{cases} \frac{r_1 \bar{r}_2}{\alpha^2}, & \text{if } (r_1, r_2) \in \Delta_1 \\ \left(\frac{1-2r_2}{2\alpha}\right)^2, & \text{if } (r_1, r_2) \in \Delta_2 \\ 1, & \text{if } (r_1, r_2) \in \Delta_3 \\ 0, & \text{if } (r_1, r_2) \in \Delta_4 \end{cases} \tag{3}$$

with $\bar{r}_i := \alpha - r_j$ for $i = 1, 2$ and $j = 3 - i$. The value φ_2 of Miner 2 is analogous and the oceanic value Φ is simply equal to $\Phi = 1 - \varphi_1 - \varphi_2$.

The interpretation of the values in the extreme regions Δ_3 and Δ_4 is straightforward. In Δ_3, miner 1 controls more than 50% of the resources and hence, has absolute power over the blockchain. This implies that her value is equal to 1 and consequently, the value for both miner 2 and the oceanic miners is 0. Region Δ_4 is similar. The interesting cases arise whenever $(r_1, r_2) \in \Delta_2$, i.e., when the major miners and the ocean, each control less than 50% of the resources, or $(r_1, r_2) \in \Delta_1$, i.e., when the resources controlled by the ocean account for more than half of the total resources. This case is also referred to as the *interior case* in the original paper. Some instantiations in regions Δ_1 and Δ_2 are presented in Table 2.

An indicative observation—which does not aim to an exhaustive analysis of the above measurements—is that the values and the ratios unveil disparities between *shares* and actual *influence* or *power* of the participating entities. For example, there are instances, as in the (40, 9, 51)-configuration (first row in Δ_1), in which a major miners' ratio is larger than the ratio of oceanic players. This imbalance generates a motive for oceanic players to merge with that miner to increase the power of their individual resources. Equivalently, the large miner has an increased influence to attract resources from the ocean. The picture is totally different in the (40, 40, 20)-configuration (third row of Δ_2), in which the competition between the major miners raises the value of resources owned by the ocenic players. Both cases can be con-

Table 2 Resources, values and values per unit of resource for various configurations in the Δ_1 and Δ_2 regions. The resources $r_i, i = 1, 2$ are selected arbitrarily, $\alpha = 1 - r_1 - r_2$, the values $\varphi, i = 1, 2$ and Φ are given by (3) and the ratios $\upsilon_i, i = 1, 2$ and υ_{oc} by (2)

	Resources %			Values %			Ratios		
	r_1	r_2	α	φ_1	φ_2	Φ	υ_1	υ_2	υ_{oc}
Δ_1: Interior game	40	9	51	65	4	31	1.62	0.42	0.62
	30	19	51	37	15	48	1.23	0.81	0.94
	25	24	51	26	24	50	1.04	1.00	0.98
Δ_2: Balance of power	35	20	45	44	11	44	1.27	0.56	0.99
	40	30	30	44	11	44	1.11	0.37	1.48
	40	40	20	25	25	50	0.63	0.63	2.5

trasted to the stability in the (25, 24, 51)-configuration, in which all 3 ratios are approximately equal to 1.

Yet, as argued in [27], the interpretation of values should be done with caution and only in addition to complementary analytical tools. This is because values do not take into account qualitative factors such as ethical commitments, operational constraints or other kinds of incentives.

3 Individual Mining Is Not Stable

A direct outcome of applying the model of Oceanic Games in the blockchain context is the next result due to [27]. Both parts of Theorem 1 make critical use of the assumption that the majority of mining resources is controlled by oceanic players. Their proof relies on a recursion in the number m of major miners and can be found in [27]. Here, we will focus on the interpretation of Theorem 1 and its application in blockchain context.

Theorem 1 ([27]) *Let* $\Gamma = [0.5; r_1, r_2, \dots, r_m; \alpha]$ *with* $M = \{1, 2, \dots, m\}$ *be a blockchain oceanic game, such that* $r(M) < 0.5 \le \alpha$, *i.e., such that the majority of mining resources is controlled by individual (oceanic) miners. Then*

(a) *The value* φ_i *of any major player* $i \in M$ *in* Γ *is given by*

$$\varphi_i = \frac{r_i}{\alpha^m} \sum_{S \subseteq M - \{i\}} \left[c_s \prod_{j \in S} r_j \prod_{k \notin S} (\alpha - r_k) \right]$$

where $c_s := s! \left[\frac{1}{s!} - \frac{1}{(s-1)!} + \cdots + (-1)^s \right]$ *and* $s := |S|$ *is the number of major miners in* S. *The oceanic value* Φ *is equal to* $\Phi = 1 - \sum_{i \in M} \varphi_i$.

(b) If $\Gamma^+ = [0.5; r_1, r_2, \ldots, r_m, r_{m+1}; \alpha]$ *for some* $r_{m+1} > 0$*, and* $\Phi^+, \varphi_i^+, i = 1, \ldots, m+1$ *are the values in* Γ^+*, then*

$$\varphi_{m+1}^+ / r_{m+1} = \Phi / \alpha$$

or equivalently, $v_{m+1}^+ = v_{oc}$.

Interpretation of Theorem 1. Statement (a) of Theorem 1 is an analytical result which yields the exact formula to compute the values of the major miners and the ocean. Its usefulness will become aparent in the applications. Statement (b) carries more intuition. It states that the value-per-unit of resource of a miner entering Γ is equal to the oceanic value-per-unit of resource in Γ. One possible interpretation, also supported by [27], is that this provides a stability argument in favor of decentralization, in the sense that there is no incentive for the formation of a "cartel" or a mining pool, *provided that* the size of the ocean is big enough, i.e., provided that the ocean controls the majority of the resources.[5]

However, as we will see in the following applications, this picture is misleading and decentralization is actually not stable. In practice, the oceanic value per unit of resource in Γ^+ can go below the value per unit of resource of the crystallizing or newly entering entity. Hence, *given that* a set of miners can coordinate their actions, then it may be beneficial for them to either crystallize out of the ocean or to acquire exogenous resources and form in both cases a single mining entity.

3.1 Applications of Theorem 1

The above interpretations of Theorem 1 are illustrated via the simulation of two representative scenarios. In both cases, we assess the stability of initial distributions of mining resources, in which the majority of resources is controlled by the ocean. This is achieved by comparing the oceanic value per unit of resource to the value per unit of resource of the same miners when acting as single entity.

I. Crystallization out of the ocean: In the first scenario, we consider an instance of the blockchain oceanic game in which all resources are initially controlled by oceanic players. This is described by the game $\Gamma = [50\%; \alpha = 100]$ and $M = \emptyset$. Then, we simulate a gradual formation of a single mining entity by the process of *crystallization* out of the ocean. This is captured by a sequence of games (instances) $\Gamma^+ = [50\%; r_1; \alpha]$ with $0 < r_1 < 50$ and $\alpha = 100 - r_1$. For each instance, we calculate the value per unit of resource of the single entity that is forming out of the ocean and compare it with the value per unit of oceanic resource. The results are shown in Fig. 3.

It is apparent that v_1 is higher than v_{oc} even for arbitrarily low values of r_1 and that the difference is increasing in the percentage of crystallized resources. This

[5]This statement actually holds for any quota $q \in (0, 1)$ and not only for $q = 0.5$ as formulated here.

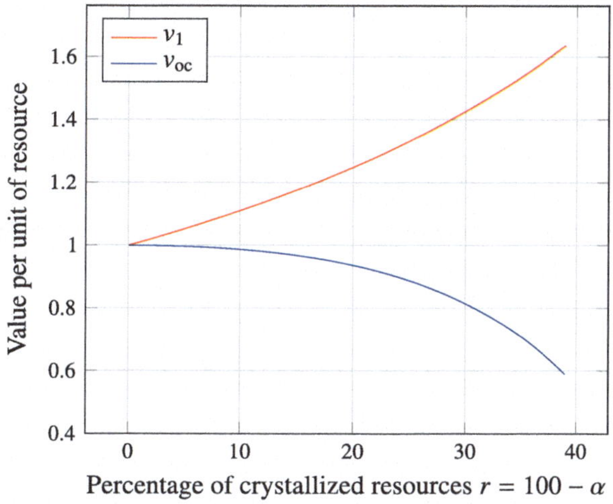

Fig. 3 The value per unit of resource of a single entity, $m = 1$, that is forming by mergers (crystallization) out of the ocean (red line) and the value per unit of oceanic resource (blue line). The total percentage of resources that is controlled by the crystallizing entity is shown in the horizontal axis

uncovers a motive for coalition formations and merging between miners, even if the initial distribution is perfectly decentralized. Further simulations (not shown here) demonstrate that the same picture continues to hold even if $M \neq \emptyset$, as long as $\alpha > 50\%$ and no single miner in M holds a percentage close to 50%. If there exists a "large" miner $i \in M$ with, e.g., $r_i > 40\%$, then the oceanic players may be disincentivized to collude. However, this is only a semblance of stability, since in this case, oceanic miners have an incentive to merge with the "large" miner.

II. Acquisition of exogenous resources: In the second scenario, we consider miners who are acquiring exogenous resources to enter the mining process. We assume that these miners can either enter the ocean and mine individually or collude and form a single mining entity. We want to compare the value per unit of resource in these two cases. Formally, we denote the current distribution of resources by $\Gamma = [50\%; r_1, r_2, \ldots, r_m; \alpha]$ with $\alpha > 50\%$ and the total mining resources of the new entities by w. We want to compare

- $v^+_{m+1} := \varphi^+_{m+1}/w$ in the game $\Gamma^+ := [50\%; r_1, r_2, \ldots, r_m, w; \alpha]$ to
- $v^{\text{o}}_{\text{oc}} := \Phi^{\text{o}}/(\alpha + w)$ in the game $\Gamma^{\text{o}} := [50\%; r_1, r_2, \ldots, r_m; \alpha + w]$.

The game Γ^+ describes the instance in which the entering miners merge in a single mining entity and the game Γ^{o} the instance in which the entering miners become part of the ocean and mine individually. We assume that initially, the majority of resources is controlled by oceanic players and that there exist two other major mining entities. It turns out that the share of mining resources of the other major entities influences the incentives of the entering miners. To see this, we consider two cases.

Case 1: Let $\Gamma = [50\%; 6, 4; 90]$, so that $\Gamma^+ = [50\%; 6, 4, w; 90]$ and $\Gamma^{\text{o}} = [50\%; 6, 4; 90 + w]$ for any $w > 0$. As shown in Fig. 4, in this case, both major miners are not large enough to create entry barriers for the third entity and $v^+_{m+1} > v^{\text{o}}_{\text{oc}}$

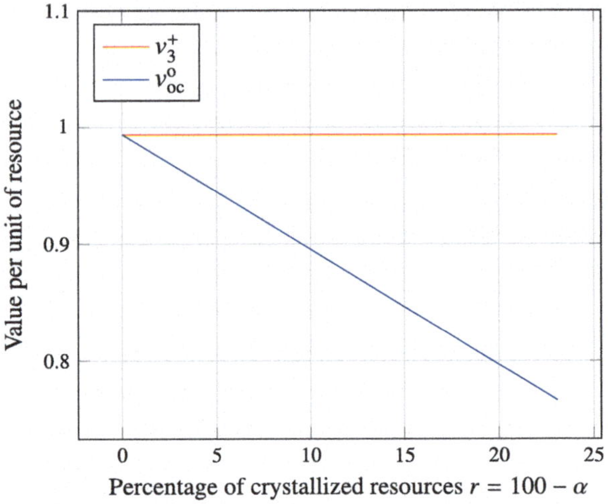

Fig. 4 The value per unit of resource of the entering miners when they enter as a single entity (red line) and their value per unit of resource when they enter as individual oceanic miners (blue line). The additional resources are shown as a percentage of the total resources in the horizontal axis

for any $w > 0$. In agreement with Theorem 1(b), the value per unit of resource, v_3^+, of the new miners when they enter as a single entity is equal to the oceanic value v_{oc} in the initial game Γ (red line). According to [27] this implies that "there is no incentive for a new entity to form". However, this only says half the truth. As we can see by the blue line, if the newly entering miners enter the ocean as individual miners, then their value per unit of resource will be lower compared to the case in which they collude. Hence, *given that* a group of entering miners are capable to coordinate their actions, then they are better off if they enter as a single entity than as oceanic players.

Case 2: Let $\Gamma = [50\%; 55, 5; 90]$, so that $\Gamma^+ = [50\%; 55, 5, w; 90]$ and $\Gamma^o = [50\%; 55, 5; 90 + w]$ for any $w > 0$. As shown in Fig. 5, in this case, the presence of major miner 1 seems to create a disincentive for a forming coalition and $v_{m+1}^+ < v_{oc}^o$ for any $w > 0$ such that $w + 90 < 50\%$. The resulting picture shows that we cannot generalize the outcome of the previous case. In particular, we conclude that whether the entering miners have an incentive to form a single entity or to join the ocean as individual miners, may depend on the actual distribution of resources among the existing major miners and the ocean. However, this is only a semblance of stability, stemming from an already centralized initial distribution ($r_1 > 33\%$). In this case, oceanic miners actually have a stronger incentive to merge with miner 1 instead of forming a new entity.

The previous simulations create an inconclusive picture. In general, the incentives for miners to merge seem to depend on the current distribution of resources. Since blockchain mining is a dynamical system that evolves over time, they suggest that even if the blockchain starts from a sufficiently decentralized point, then it is unlikely to remain decentralized also in the future or equivalently that decentralization creates a *negative feedback loop*, [19, 36]. The dynamics of the coalition formation process and the entry barriers resemble these of conventional economic markets of either

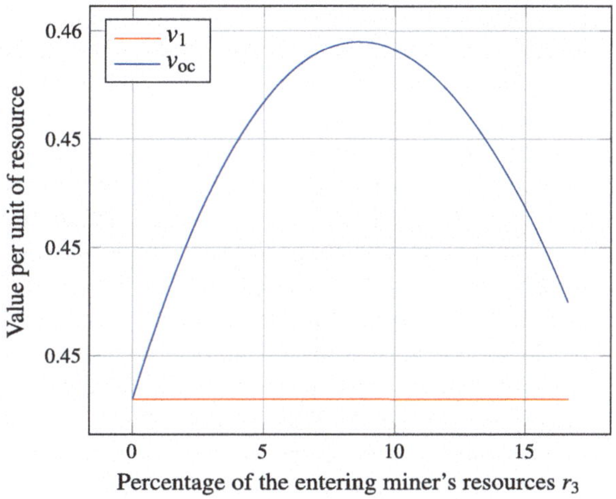

Fig. 5 The value per unit of resource of the entering miners when they enter as a single entity (red line) and their value per unit of resource when they enter as individual oceanic miners (blue line). The additional resources are shown as a percentage of the total resources in the horizontal axis

perfect or oligopolistic competition. These findings provide an alternative perspective to cryptocurrency mining along with [1], and suggest the need for further research in this direction.

4 General Issues, Research Perspectives and Limitations

The application of the oceanic-game model in blockchain mining opens new research perspectives but also has its own limitations. Beyond the insight from existing results, a complete model needs to account for the additional challenges and address the questions that are specific to the blockchain context. In the following discussion, we raise such relevant issues, discuss their connection and research possibilities via the current model and identify potential limitations.

Cryptocurrencies as Resources: The difference between PoW and PoS in terms of their resources—computational power versus native coins—has a direct impact on both the mining process and the value of the underlying cryptocurrency. When coins are used as mining resources (PoS protocols), their value depends on their distribution among existing miners, their availability for prospective miners and the returns (profits) from mining. This in contrast to PoW protocols, in which the price of the resources—e.g., hardware and electricity—is not tied to the price of the underlying cryptocurrency.

Resource Acquisition & Entry Barriers: The above suggest that the nature of protocol resources may also generate different entry barriers. In PoS, the acquisition of protocol resources, i.e., coins, by prospective nodes depends on the willingness of current owners to exchange their coins and the way that new coins are minted. Different configurations may lead to high entry barriers and centralization. In

PoW, computational power is essentially unlimited and acquisition of additional resources is independent of the underlying cryptocurrency. This implies lower entry barriers but also more frequent changes in the configuration (distribution) of resources among miners.

In particular, the current cost of acquiring enough computational power to control the majority in the Bitcoin and Ethereum PoW-blockchains is estimated at 1.5 billion US Dollars [6]. This amount is well within the budget of several physical or legal entities worldwide. Moreover, it is *independent* of the value of the underlying cryptocurrency and depends only on the size of the network and the hardware and electricity costs. With this in mind, it is natural to ask: how stable are PoW blockchains against arbitrary authorities able to acquire the majority of resources? How relevant are these questions to the current distribution of resources and how do they translate in the PoS setting?

In this context, further work on blockchain oceanic games can aid the community to raise and study questions about investment in cryptocurrencies. When viewed not only as assets but also as means to gain power in the mining process and the governance of a blockchain, cryptocurrencies fit to the current perspective and their mechanics can be better understood.

Mathematical Modelling: From a mathematical perspective, oceanic games bridge the gap between atomic and non-atomic congestion games [23]. Yet, the use of *values* instead of equlibria to study real settings has its own limitations [27, 32]. This is mainly due to the probabilistic derivation of values, which ignores qualitative aspects such as ethical commitments, preferences or any other motives of the participating agents. However, despite these limitations, if properly interpreted, values can become a powerful tool in the analysis of strategic interactions. In an immediate direction, they can be used to rethink the notion of *blockchain fairness* or *equitability*, which is currently based on the theoretically tentative premise that *one unit of resource—one vote* also implies fairness [13, 35].

5 Conclusions

In this paper, we employed the concept of Oceanic Games [27], to model and study strategic interactions in blockchain mining. Oceanic Games have been used in conventional economics to analyze decision making in corporate settings with a small number of major players—shareholders or here, mining pools—and a continuum of minor, individually insignificant players, called the *ocean*. This stream of literature focuses on the measurement of the value per miner and per unit of resource for each miner given a distribution of resources. Values are then interpreted as strategic components in decisions related to resource acquisition, mergers and coalition formations and offer an alternative perspective to the common equilibration models.

An immediate implication of existing results was that given a sufficiently large initial distribution of resources, there are *incentives* both for active and for newly entering miners to merge (form cartels or coalitions) and act as single entities. These observations provide an alternative justification of the observed centralization and concentration of power in the mining process of the major cryptocurrencies. Contrary to common perceptions, they amount to the existence of a negative feedback loop in terms of decentralization as a core ingredient in permissionless blockchain philosophy, [17], and reveal the need for futher research in this direction. In a general discussion, we identified critical issues related to resource acquisition, entry barriers and centralization risks in blockchain mining and formulated relevant questions that may be answered by further exploration of the present model. These findings can be placed in the broader context of governance and long-term sustainability for pemissionless blockchains.

References

1. Arnosti, N., Weinberg, S.M.: Bitcoin: A natural oligopoly. In: 10th Innovations in Theoretical Computer Science Conference, ITCS 2019, pp. 5:1–5:1. San Diego, USA (2019). https://doi.org/10.4230/LIPIcs.ITCS.2019.5
2. Aumann, R.J.: Markets with a continuum of traders. Econometrica **32**(1/2), 39–50 (1964)
3. Badertscher, C., Garay, J., Maurer, U., Tschudi, D., Zikas, V.: But why does it work? a rational protocol design treatment of bitcoin. In: Nielsen, J.B., Rijmen, V. (eds.) Advances in Cryptology—EUROCRYPT 2018, pp. 34–65. Springer International Publishing, Cham (2018)
4. Badertscher, C., Gaži, P., Kiayias, A., Russell, A., Zikas, V.: Ouroboros genesis: Composable proof-of-stake blockchains with dynamic availability. In: Proceedings of the 2018 ACM SIGSAC Conference on Computer and Communications Security, pp. 913–930. CCS '18, ACM, USA (2018). https://doi.org/10.1145/3243734.3243848
5. Bentov, I., Gabizon, A., Mizrahi, A.: Cryptocurrencies Without Proof of Work. In: Clark, J., Meiklejohn, S., Ryan, P.Y., Wallach, D., Brenner, M., Rohloff, K. (eds.) Financial Cryptography and Data Security, pp. 142–157. Springer, Heidelberg (2016)
6. Bonneau, J.: Hostile blockchain takeovers (short paper). In: Proceedings of the 5th IFCA Workshop on Bitcoin and Blockchain Research (2018)
7. Bonneau, J., Felten, E., Goldfeder, S., Kroll, J., Narayanan, A.: Why Buy When You Can Rent? Bribery Attacks on Bitcoin-Style Consensus. In: Financial Cryptography Workshops (2016)
8. Bonneau, J., Miller, A., Clark, J., Narayanan, A., Kroll, J.A., Felten, E.W.: SoK: Research perspectives and challenges for bitcoin and cryptocurrencies. In: 2015 IEEE Symposium on Security and Privacy, pp. 104–121 (2015). https://doi.org/10.1109/SP.2015.14
9. Brünjes, L., Kiayias, A., Koutsoupias, E., Stouka, A.P.: Reward Sharing Schemes for Stake Pools (2018). arXiv:1807.11218
10. Buterin, V., Reijsbergen, D., Leonardos, S., Piliouras, G.: Incentives in Ethereum's hybrid casper protocol. In: ICBC 2019. Seoul, Korea (2019). https://doi.org/10.1109/BLOC.2019.8751241
11. Eyal, I., Sirer, E.: Majority is not enough: bitcoin mining is vulnerable. In: Christin, N., Safavi-Naini, R. (eds.) Financial Cryptography and Data Security, pp. 436–454. Springer, Heidelberg (2014)
12. Eyal, I.: The Miner's Dilemma. In: Proceedings of the 2015 IEEE Symposium on Security and Privacy. SP '15, pp. 89–103. IEEE Computer Society, USA (2015). https://doi.org/10.1109/SP.2015.13
13. Fanti, G., Kogan, L., Oh, S., Ruan, K., Viswanath, P., Wang, G.: Compounding of Wealth in Proof-of-Stake Cryptocurrencies (2018). ArXiv e-prints

14. Fisch, B., Pass, R., Shelat, A.: Socially optimal mining pools. In: R. Devanur, N., Lu, P. (eds.) Web and Internet Economics, pp. 205–218. Springer International Publishing, Cham (2017)
15. Garay, J., Kiayias, A., Leonardos, N.: The Bitcoin backbone protocol: analysis and applications. In: Annual International Conference on the Theory and Applications of Cryptographic Techniques, pp. 281–310. Springer, Berlin (2015)
16. Garay, J., Kiayias, A., Leonardos, N.: The Bitcoin backbone protocol with chains of variable difficulty. In: Katz, J., Shacham, H. (eds.) Advances in Cryptology - CRYPTO 2017, pp. 291–323. Springer International Publishing, Cham (2017)
17. Garay, J.A., Kiayias, A., Leonardos, N., Panagiotakos, G.: Bootstrapping the blockchain, with applications to consensus and fast PKI setup. In: Public Key Cryptography (2018). https://eprint.iacr.org/2016/991
18. Gilad, Y., Hemo, R., Micali, S., Vlachos, G., Zeldovich, N.: Algorand: Scaling byzantine agreements for cryptocurrencies. In: Proceedings of the 26th Symposium on Operating Systems Principles. SOSP '17, pp. 51–68. ACM, USA (2017). https://doi.org/10.1145/3132747.3132757
19. Hauert, C., De Monte, S., Hofbauer, J., Sigmund, K.: Volunteering as Red Queen Mechanism for Cooperation in Public Goods Games. Science **296**(5570), 1129–1132 (2002). https://doi.org/10.1126/science.1070582
20. Johnson, B., Laszka, A., Grossklags, J., Vasek, M., Moore, T.: Game-theoretic analysis of DDoS attacks against bitcoin mining pools. In: Böhme, R., Brenner, M., Moore, T., Smith, M. (eds.) Financial Cryptography and Data Security, pp. 72–86. Springer, Heidelberg (2014)
21. Kiayias, A., Russell, A., David, B., Oliynykov, R.: ouroboros: A provably secure proof-of-stake blockchain protocol. In: Katz, J., Shacham, H. (eds.) Advances in Cryptology –CRYPTO 2017. pp. 357–388. Springer International Publishing, Cham (2017). https://doi.org/10.1007/978-3-319-63688-7_12
22. Kiayias, A., Koutsoupias, E., Kyropoulou, M., Tselekounis, Y.: Blockchain mining games. In: Proceedings of the 2016 ACM Conference on Economics and Computation, pp. 365–382. ACM, USA (2016)
23. Kleinberg, R., Piliouras, G., Tardos, E.: Multiplicative updates outperform generic no-regret learning in congestion games: extended abstract. In: Proceedings of the Forty-first Annual ACM Symposium on Theory of Computing. STOC '09, pp. 533–542. ACM, USA (2009). https://doi.org/10.1145/1536414.1536487
24. Laszka, A., Johnson, B., Grossklags, J.: When Bitcoin mining pools run dry. In: Brenner, M., Christin, N., Johnson, B., Rohloff, K. (eds.) Financial Cryptography and Data Security, pp. 63–77. Springer, Heidelberg (2015)
25. Leonardos, S., Reijsbergen, D., Piliouras, G.: Weighted voting on the blockchain: Improving consensus in proof of stake protocols. In: ICBC 2019. Seoul, Korea (2019). https://doi.org/10.1109/BLOC.2019.8751290
26. Lewenberg, Y., Bachrach, Y., Sompolinsky, Y., Zohar, A., Rosenschein, J.S.: Bitcoin mining pools: a cooperative game theoretic analysis. In: Proceedings of the 2015 International Conference on Autonomous Agents and Multiagent Systems. AAMAS '15, pp. 919–927. International Foundation for Autonomous Agents and Multiagent Systems, Richland, SC (2015)
27. Milnor, J.W., Shapley, L.S.: Values of Large Games II: Oceanic Games. Mathematics of Operations Research **3**(4), 290–307 (1978). https://doi.org/10.1287/moor.3.4.290
28. Nakamoto, S.: Bitcoin: A Peer-to-Peer Electronic Cash System (2008). https://bitcoin.org/bitcoin.pdf. [Accessed: 14 Nov 2018]
29. Nayak, K., Kumar, S., Miller, A., Shi, E.: Stubborn mining: generalizing selfish mining and combining with an eclipse attack. In: 2016 IEEE European Symposium on Security and Privacy (EuroS P), pp. 305–320 (2016). https://doi.org/10.1109/EuroSP.2016.32
30. Pass, R., Shi, E.: Thunderella: blockchains with optimistic instant confirmation. In: Nielsen, J.B., Rijmen, V. (eds.) Advances in Cryptology—EUROCRYPT 2018, pp. 3–33. Springer International Publishing, Cham (2018). https://doi.org/10.1007/978-3-319-78375-8_1
31. Shapiro, N.Z., Shapley, L.S.: Values of large games, I: a limit theorem. Math. Oper. Res. **3**(1), 1–9 (1978). https://doi.org/10.1287/moor.3.1.1

32. Shapley, L.S.: A value for n-person games. In: Roth, A.E. (ed.) The shapley value: essays in Honor of Lloyd S. Shapley, pp. 31–40. Cambridge University Press, Cambridge (1988). https://doi.org/10.1017/CBO9780511528446.003
33. Shitovitz, B.: Oligopoly in markets with a continuum of traders. Econometrica **41**(3), 467–501 (1973)
34. Sompolinsky, Y., Zohar, A.: Bitcoin's Underlying Incentives. Commun. ACM **61**(3), 46–53 (2018). https://doi.org/10.1145/3152481
35. Wong, T.: An application of game theory to corporate governance. Omega **17**(1), 59–67 (1989). https://doi.org/10.1016/0305-0483(89)90021-2
36. Zeigler, B., Praehofer, H., Kim, T.: Theory of Modeling and Simulation: Integrating Discrete Event and Continuous Complex Dynamic Systems, 2nd edn. Academic Press (2000)

Smart Contract-Driven Mechanism Design to Mitigate Information Diffusion in Social Networks

Arinjita Paul, Vorapong Suppakitpaisarn and C. Pandu Rangan

Abstract This paper presents a new direction in privacy preserving techniques for social networks based on consensus-driven blockchain and mechanism design principles. Privacy problem is among the class of the most important and fundamental problems in social networks. The most commonly accepted privacy solution is to incorporate a perfect data privacy policy and central system, which inherently lacks transparency and trust. All existing privacy techniques deny undesired users access to the information directly, but, in reality, the information may be forwarded to them from other users who possess the information. Our user-controlled privacy mechanism aims to control such data dissemination using simple game theoretic concepts combined with blockchain technology. Our mechanism applies to DAG structured networks (directed acyclic graphs), and our reward policy incentivizes the receivers if they do not diffuse the message in the network. We establish blockchain powered smart contracts to enable the flow of incentives in the system, without the involvement of a trusted third party. The owner of the message has to pay for the rewards, but our mechanism makes sure that the payment is minimum. In fact, the owner will have more utility when he/she pays. Our mechanism satisfies the necessary constraints of mechanism design, namely individual rationality, incentive compatibility, and weakly budget balance while ensuring privacy.

A. Paul (✉) · C. P. Rangan
Department of Computer Science and Engineering, IIT Madras, Chennai, India
e-mail: arinjita@cse.iitm.ac.in

C. P. Rangan
e-mail: prangan@cse.iitm.ac.in

V. Suppakitpaisarn
Graduate School of Information Science and Technology,
The University of Tokyo, Tokyo, Japan
e-mail: vorapong@is.s.u-tokyo.ac.jp

P. Pardalos et al. (eds.), *Mathematical Research for Blockchain Economy*,
Springer Proceedings in Business and Economics,
https://doi.org/10.1007/978-3-030-37110-4_14

1 Introduction

The advent of blockchain technology has led to a paradigm shift from centralised to a decentralised and autonomous control. Blockchain is a decentralised verifiable public ledger which maintains records of transactions in an append-only fashion. Identical copies of the blockchain are distributed among each participating nodes in the network, and any changes to the ledger are reflected across all copies. Initially envisioned for secure transfer of decentralised digital currency [25], the technology has been extended to provide a generalised framework for implementing decentralised applications requiring trusted computing and auditability [35], such as finance, Internet of Things, governance applications, capital markets and e-health [7].

Social networking is pervasive in today's world, leading to a boom in the information economy. Social networking sites have become the preferred medium of information sharing with peers by means of purchases, queries, conversations and other related activities. However, such a popularity has been accompanied by growing concerns for privacy of its users [34]. Such sites form a database of personal data that holds substantial economic value, serving as a hotbed for potential marketing networks, malware, spam, illegal earnings and several other attacks [22, 34] on the Internet. Leakage of critical data such as medical health records and business information regarded as a business asset, could lead to dire consequences such as identity theft, financial loss, harassment and fraud [12]. Social network privacy or informational privacy is still in its infancy, with no well-defined security model. In 2015, medical data of 78.8 million patients, nearly a quarter of the U.S. population, were stolen by a hack on the insurance corporation Anthem as a result of weak policy enforcement and security systems [7].

To address these concerns, social network service providers have developed privacy policies and features [15] to balance the trade-off between privacy threats and data sharing. Note that such privacy measures are traditionally supported by centralized systems, which lack trust and transparency, as evident from the recent breaches in privacy reported in [11, 13].

1.1 Our Contribution

In an attempt to shift from a centralised privacy system to a user-controlled approach, we propose a new direction that employs mechanism design theory combined with blockchain technology to ensure privacy of sensitive data in social networks while improving societal welfare, which hitherto has not been explored in the literature. We propose a novel mechanism, called the Social Network Privacy Mechanism (SNPM), that incentivizes the participants for not diffusing the private information into the network. We leverage blockchain enabled smart contracts [3] as a decentralised approach to automatize the incentive system and tackle trust issues in our privacy mechanism. We prove that our mechanism satisfies all the required properties for a mechanism to function. Those properties are individual rationality, incentive compatibility and weakly budget balance.

We consider the social network as a directed acyclic graph (DAG). Several algorithms for general graphs are obtained from algorithms for DAGs such as [29]. There exist several results that consider social networks with *time label* [17]. Again, temporal social networks can be considered as DAGs [18]. Although it is not straightforward to apply our mechanism to periodic networks, we strongly believe that it could be a possible extension of our mechanism to general graph structures.

1.2 Example

Our mechanism is demonstrated in Fig. 1. A user m wishes to propagate a message only to a set of his neighbours and promises a reward to each receiver who does not propagate the message in the network. To prevent message dissemination, m has a reward amount to incentivize his neighbours (agents a and b here) for not propagating the message. As shown in the figure, user m rewards neighbour b with an amount $R_e = 35$ as an incentive to not propagate his message to user e. Again, user e could bribe user b with a value $B_e = 28$ to acquire the message from b. Since $R_e > B_e$, user b refrains from propagating the message to e, as indicated by red arrows. Our mechanism displays how the reward value is computed by each user in the network, where all financial transactions are performed by a smart contract, to restrict users from message propagation, and demonstrates how equilibrium is attained while preserving message privacy along with the necessary properties of mechanism design.

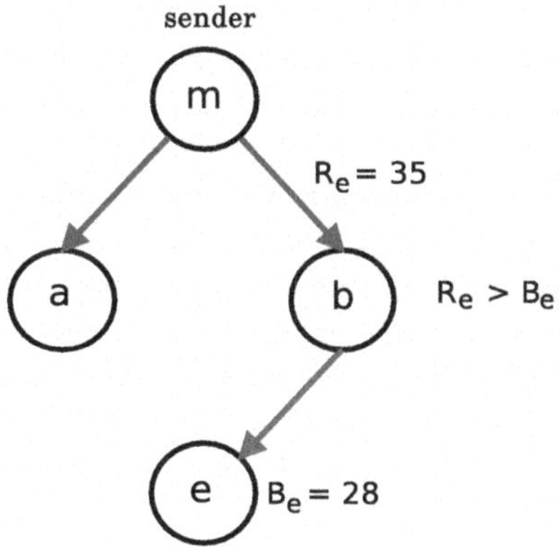

Fig. 1 A social network example to demonstrate our privacy mechanism. A user m shares private information only with users (a, b). The arrows green symbolize message transmission by m to (a, b) and red symbolizes forbidden dissemination to agent e

2 Related Works

2.1 Integrating Blockchain Technology to Social Networks

Although the blockchain paradigm was originally designed to maintain a decentralised financial ledger, it has been extended to serve other applications requiring trusted computing and auditability. Recently, several research efforts have focused on blockchain based social networks. This establishes a decentralised approach to connectivity to get rid of a centralised server, preventing any single authority from enforcing monitoring and control over the user generated data for financial incentives. Some examples of decentralised social networks include Akasha [1], Diaspora [8] and Steemit [30] among others. Such principles of decentralization has also been applied for managing large data, such as Ancile [7] and MedRec [2] for electronic health records (EHR) management in a medical network and for personal data [35] management. Note that, one downside of the blockchain technology is that it is resource intensive, and hence scalability is an issue for large scale systems. All existing results rely on the distributed ledger mechanism and external regulations such as the HIPAA privacy rule to address security of individual data. In this paper, privacy of user data is maintained using simple mechanism design principles and the incentives are managed using blockchain. Unlike the previous works stated, the transactions maintained in our ledger are purely financial, to only regulate the financial incentives in the system.

2.2 Anonymization of Social Networks

Social Network Anonymization is a countermeasure of linking attack, where undesired users can infer protected information from published data. Such an attack is prevented by removal or perturbation of certain information, while satisfying privacy notions. In relational databases, the most commonly accepted privacy notions include k-anonymity [31], ℓ-diversity [23] or t-closeness [21]. In the context of social network privacy, such notions are extended to concepts such as k-isomorphism [5]. Several graph modification techniques such as graph perturbation [14] and clustering approaches [4] have been introduced, that meet privacy notions for social networks.

Social network anonymization usually addresses privacy problems arising from data publication. On the other hand, we consider a different dimension of information privacy in this work, i.e., preventing private data diffusion in social networks.

2.3 *Mechanism Design Towards Social Choices in Networks*

In addition to the preference of outcomes of a mechanism (winner in an auction), users may also be concerned with what private information gets leaked to others (the valuation of the auctioned item). The latter notion of privacy has been addressed in the literature using techniques of differential privacy [27, 28]. Informally, differential privacy [10] captures the fact that a change in a single agent's input has too small an effect to jeopardize the privacy and learn any information about the agent from the outcome of a joint computation. Note that, differential privacy offers a compelling second-best solution concept when the exact dominant strategy truthful mechanism is not known [28] as it offers *approximate-truthfulness* [9].

Li et al. [20] applies mechanism design theory to the auction design problem for a seller to sell a commodity in a social network. While their work focuses on increasing the number of users that know the auction information and maximizing the auction bid, our work aims to minimize the number of participants that know the private information. One might think that our mechanism is analogous to the mechanism due to Li et al. However, because of the difference in objectives, the two mechanisms do not share any similarity with each other.

3 Preliminaries

3.1 *Blockchain*

Blockchain is a distributed ledger technology typically managed by a peer-to-peer network. Its non-trusting members record digital transactions into the shared immutable ledger in a verifiable manner without the need of a centralised regulator. Blockchain implements the concept of *mining* and *proofs* inorder to reach a consensus on the transaction ordering in a decentralized fashion. *Miners* are a subset of the network participants whose role is to validate transactions broadcasted into the network and append these transactions grouped into a *block* to the blockchain. To this end, they fiercely compete with one another to solve difficult computational problems, and are rewarded (usually monetary) for their service. *Proofs* determine which miner's block will be appended next to the blockchain, such as proof-of-work and proof-of-stake. Cryptocurrencies such as Bitcoin [25] and Ethereum [3] are built atop such a technology, wherein the network members run distributed consensus protocols. Recently, there has been an increasing interest to exploit the technology to develop applications beyond digital currencies requiring tamper-proof network consensus. Blockchain has attracted the interest of stakeholders across a wide range of industries owing to its decentralised approach towards providing trust and integrity in the network, such as healthcare, real estate, finance, cloud storage, governance applications among others.

3.2 Smart Contracts

Introduced in 1994 by Nick Szabo, a smart contract is "a computerized transaction protocol that executes the terms of a contract" [32]. It is a user defined software executed by a network of mutually distrustful parties, and has received much attention in the context of blockchains. Its correct execution is automatically enforced without the arbitration of any central authority, and stores its result on the blockchain. One such example of smart contracts are Ethereum [3], which builds a Turing-complete instruction set to allow smart contract programming into the blockchain, and records the contract states on the blockchain. Depending on the intended application, smart contracts could be used towards financial, notarial or game-based applications among others. Since such scripts are tamper resilient and their actions are publicly visible, they are appealing in scenarios that require transfer of money to respect certain agreed rules. We regard such a feature offered by smart contracts as an important property to achieve financial fairness in our privacy mechanism.

3.3 Mechanism Design

Mechanism design is a fundamental concept in economics and AI [26]. It is an art of creating economic interactions with respect to a preferable outcome of the game induced by the mechanism. We closely follow the mechanism design framework in [24]. We consider a social network consisting of n persons or agents represented by a set N, where each agent is indexed by $i \in \{1, \dots, n\}$. Every agent $i \in N$ must report its action $a_i \in A_i$ for the public decision, where A_i denotes the complete action space of an agent i possible towards social welfare. Let $a = (a_1, a_2, \dots a_n)$ be a vector that denotes the action profile of all the agents $i \in N$, and A denotes the complete action space for all agents in the network. We use the following notation N_{-i} to denote the set $N \setminus \{i\}$ which is the set of all agents except agent i, and the notation a_{-i} to denote the action profile of all agents except agent i.

In the auction setting, every agents bid for a number of objects. For simplicity, in this paper, we will assume that the number of objects is one. Every agent $i \in N$ has a value of the object $v_i \in \mathbb{R}$ that indicates his willingness or valuation on that object. The agents then take an action to report how much they want to pay for the bid. By that, the action set A_i denotes the set of all possible report values. Let $\pi_i(a)$ be a decision function of the mechanism where $\pi_i(a) = 1$ when the agent i is a winner who can receive the object and $\pi_i(a) = 0$ otherwise. The winner i will then have to pay an amount equal to p_i for the object to the auctioneer. In the most well-known Vickrey mechanism [33], the winner is the agent that report the largest value, i.e. $pi_i(a) = 1$ if and only if $a_i = \max_{j \in N} a_j$, and the price that she has to pay is equal to the second largest value, i.e. $p_i = \max_{j \in N_{-i}} a_j$. The utility of the agent i, denoted by $u_i(a)$ is then computed as $u_i(a) = \pi_i(a) \cdot (v_i - p_i)$.

Given the basic definitions, we formalize the desirable three criteria for evaluating a mechanism, which are *individual rationality*, *incentive compatibility* and *weakly-budget balanced*.

Since every agent a_i is rational, their action a_i is in the selfish interest to maximize their individual utilities $u_i(a)$. She might bid with the values larger or smaller than her evaluation on the object. That usually lead to a smaller social welfare $\sum_{i=1}^{n} u_i(a)$. If a mechanism is individually rational, the agent can be sure that reporting a honest value, $a_i = v_i$ in the auction mechanism, never lead to a negative utility.

Definition 1 A mechanism is individually rational if $u_i(v_i, a'_{-i}) \geq 0$ for all $i \in N$, and $a'_{-i} \in A_{-i}$.

The incentive compatibility then guarantees that reporting the honest value will lead to the maximum utility.

Definition 2 A mechanism is incentive compatible if $u_i(v_i, a_{-i}) \geq u_i(a_i, a'_{-i})$ for all $i \in N$, $a_i \in A_i$ and $\{a_{-i}, a'_{-i}\} \in A_{-i}$.

Next, we will define the notion of weakly-budget balance.

Definition 3 A mechanism is weakly-budget balanced if the payment policy p does not exhibit a budget deficit for a utility profile u, i.e.,

$$\sum_{i=0}^{n} p_i \geq 0$$

Note that $\sum_{i=0}^{n} p_i$ is a sum of all agents' payment in the network. The intuition for a mechanism to be weakly-budget balanced is that, in case of a negative revenue, a payment made by agents is not covered by the payment received by the agents in the network. Besides, there is no external source to finance the mechanism to function and provide an outcome.

4 Social Network Privacy Mechanism (SNPM)

In this section, we design a mechanism to conquer the problem of privacy in social networks, which we call Social Network Privacy Mechanism (SNPM). We show that our mechanism satisfies all the necessary properties, i.e., it is individually rational, incentive-compatible and weakly budget balanced. First, we give an overview of our model, to demonstrate the setting on which we enforce social network privacy.

4.1 Our Model

In our model, each agent $i \in N = \{1, \ldots, n\}$ in the social network has a set of neighbours denoted by $d_i \subseteq N$, with whom the agents can communicate directly via the link/edge. An agent termed as messenger $m \in N$ has a private message and wishes to propagate the message to only a selected set of neighbours and discourage them from further propagating to other agents in the network. Every agent in the network is oblivious of the presence of other agents except his neighbours. In our model, the messenger is not aware of the network structure beyond its neighbours.

We consider that the information diffusion flow in the network forms a direct acyclic graph (DAG). Such an assumption is reasonable, evident from the existing results in the social network literature [19], and extensions of algorithms on DAGs to general graph structures [29]. There exists algorithms for conversion of periodic graphs to DAGS [6], which could be a possible albeit challenging extension of our mechanism to incorporate generic network structures.

In order to achieve a decentralised, permanent and uncensorable mode of payment, we use blockchain systems like Ethereum [3] and NEO [16], wherein every agent joins the network (generates a unique identifier for each agent) and can access all the transactional information on blockchain. Such a system maintains a log of transactions of the agents via smart contracts. It relies on the multiple participating entities in the system to avoid a single-point-of-failure and single-point-of-breach. This makes the business model and incentive structure much robust and trusted, instead of assuming the presence of a trusted authority. Our main idea is to reward the agents who do not propagate the message. To this end, the messenger intends to reward amount values U_i for every agent $i \in N_{-m}$, allocated based on the preferences of message sharing of the messenger. All the reward and bribe transactions are automated through smart contracts.

U_i denotes the benefit of the messenger if the message is not propagated to agent i. The utility of the messenger will increase by $U_i \geq 0$ if the agent i do not receive the message. On the other hand, every agent $i \in N_{-m}$ maintains a valuation v_i for the message propagated by the messenger. The utility of the agents $i \in N_{-m}$ increases by a value $v_i \geq 0$ on acquiring the information. Agents who do not receive the message from m may bribe their neighbours with an amount $v_i' \leq v_i$ to acquire the message. Again, the messenger rewards the non-propagating nodes (via smart contracts) with a reward value $r_i \leq U_i$ for not propagating the message. Hereon, we consider all the transactions between the messenger and the agents are automated by smart contracts.

In our model, the action of every agent i is denoted as a tuple $a_i = (v_i', d_i')$. The value $v_i' \leq v_i$ is the bribe i pays to a neighbour who receives the message. The set $d_i' \subset d_i$ is the set of neighbours (descendants) to whom i spreads the message on receipt of bribe. The action space A_i is $V_i \times \mathcal{P}(d_i)$ where V_i represents the set of real number no larger than v_i and $\mathcal{P}(d_i)$ represents the power set of the neighbour set d_i.

Our decision function is represented by $\pi_i : A \rightarrow \{0, 1\}$, where $\pi_i(a) = 1$ if agent i is allocated the message due to the action a, and $\pi_i(a) = 0$ otherwise. Therefore, we denote the set $\pi = \{\pi_i\}_{i \in N_{-m}}$ as an *allocation policy* in this work.

The motive of every agent i is to acquire the message and their action a_i are in the selfish interest to maximize their individual utilities while receiving the message. Therefore, the agents could misreport their valuations (bribe) of the message. A privacy mechanism is *individual rational* if the utility of an agent reporting true valuations is not negative.

A mechanism is *incentive compatible*, if reporting the true valuations by the agents in the mechanism is a dominant strategy.

The revenue generated by the mechanism is calculated as a sum of the payment balance between the messenger and other agents in the network in an action profile. The reward sum of the messenger equals $Rev_m(a) = \sum_{i \in N_{-m}} r_i$, while every agent $i \in N_{-m}$ pays their neighbours a sum equal to $Rev_i(a) = \sum_{i \in N_{-m}} p_i$. The total revenue for an action profile must be non-negative, to avoid shortage of budget in order that the mechanism is *weakly budget balanced*. It is easy to follow that the revenue generated by our mechanism $Rev_i(a) = \sum_{i \in N} p_i = 0$. For all agents $i \in N_{-m}$, the reward value r_i paid by the messenger to the agent i and the bribe sum $v_j, \forall j \in d'_i$ is annulled by the payment p_i made by the agent i in the overall revenue, thereby resulting in a zero sum.

4.2 Our Mechanism

Given our model, we next propose our mechanism for social network privacy. An overview of our mechanism is shown in Fig. 2. We design a recursive strategy to define our privacy mechanism. We use the following notations. Let B_i denote the total bribe amount that agent $i \in N_{-m}$ can offer to its neighbour who possesses a message. Let R_i denote the total reward amount of the messenger m for an agent i,

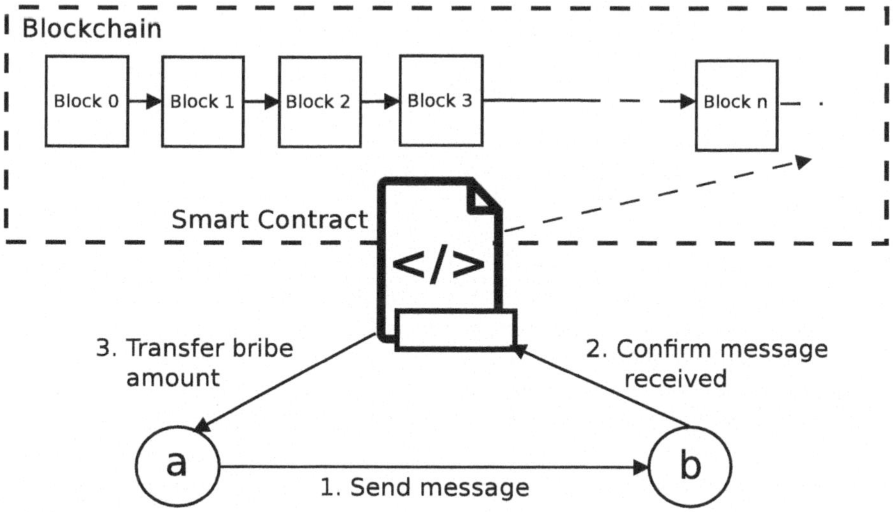

Fig. 2 Overview of SNPM. In the example, if agent a can acquire a higher amount of bribe B_a from agent b than the reward R_a, it sends the message to agent b. On receiving the message, the smart contract confirms the message receipt from b, and transfers the bribe amount from agent b to agent a and records on the blockchain

sufficient to stop i from receiving a bribe. Therefore, the total bribe i can pay for a message is the sum of the bribe received from his neighbour set d_i' along with his own bribe amount v_i. Therefore, every agent i can compute its budget as below:

$$B_i = v_i + \sum_{j \in d_i', U_j \geq v_j} (B_j + \epsilon) + \sum_{j \in d_i', U_j < v_j} (R_j).$$

The above formula defines a recursive structure for the computation of the total budget of an agent in order to bribe his neighbour possessing the message. Every agent can pay a bribe value equivalent to the bribe sum of its descendants. Also, the reward sum R_i for a messenger is the sum of the messenger's utility for all the descendants of a non-propagating agent.

The reward amount R_i of the messenger m is the collective utility of m if agent i does not receive the message (utility is U_i) and the utility of its descendants (utility is $U_j, \forall j \in d_i'$), minus the total reward value paid by m if i along with its descendant agents receive the message and do not propagate it to their descendants. We denote the collective utility as $\mathbb{U}_i$ and the cumulative reward as $\mathbb{R}_i$. Let $(\mathbb{U}_i)_{without}$ denote the collective utility of the messenger when agent i does not receive the message, and $(\mathbb{U}_i)_{with}$ to denote the same when i receives the message. Similarly, let $(\mathbb{R}_i)_{without}$ denote the collective reward for agent i when it does not receive the message, and $(\mathbb{R}_i)_{with}$ to denote the same when i receives the message. From the concept of VCG, the total reward amount of the messenger to offer agent i is computed as below:

$$\begin{aligned}
R_i &= (\mathbb{U}_i)_{without} - (\mathbb{R}_i)_{without} - ((\mathbb{U}_i)_{with} - (\mathbb{R}_i)_{with}) \\
&= (\mathbb{U}_i)_{without} - 0 - (\mathbb{U}_i)_{with} + (\mathbb{R}_i)_{with} \\
&= U_i + \sum_{j \in d_i'} (\mathbb{U}_j)_{without} + \sum_{\substack{j \in d_i', \\ U_j \geq v_j}} (B_j + \epsilon) \\
&= U_i + \sum_{\substack{j \in d_i', \\ U_j < v_j}} (R_j) + \sum_{\substack{j \in d_i', \\ U_j \geq v_j}} (B_j + \epsilon).
\end{aligned}$$

Note that, in the above derivation, $(\mathbb{U}_i)_{without}$ is the utility of the messenger when agent i does not receive the message, which is U_i. Also, $(\mathbb{R}_i)_{without} = 0$ since an agent i who does not possess the private message will not be rewarded as per the mechanism. Given the scenario, the motive of the mechanism is to convince an agent possessing the message to not accept bribe with the reward value. Naturally, for an agent i, if $R_i > B_i$, the agent is motivated to not receive any bribe from its descendants. Now we propose our mechanism based on the above definitions.

Definition 4 (*Privacy Mechanism SNPM*) Given the action profile a of the agents, the privacy mechanism SNPM is defined by an allocation policy π and a payment policy p, which are defined as follows.

The allocation policy of the privacy mechanism is defined as:

$$\pi_i(a) = \begin{cases} 1, & \text{if } B_i \geq R_i \\ 0, & \text{otherwise} \end{cases}$$

Note that, there could exist multiple agents who are allocated the message, that is, multiple agents with sufficient $B_i \geq R_i$ have the budget to bribe agents to acquire the message, which also represents the subset of users for whom m does not have enough reward to stop propagation. There exists a reward policy, which is the incentive given by the messenger to the agents who do not diffuse the message to his descendants. The reward policy of the privacy mechanism is defined as:

$$r_i(a) = \sum_{j \in d_i'} (1 - \pi_j)(B_j + \epsilon).$$

Assume that under the allocation policy, an agent i gets the message, the payment policy is defined as follows:

$$p_i = B_i - \sum_{j \in d_i'} \pi_j(a) \cdot R_j - r_i(a)$$

The privacy mechanism allocates the message to all the agents whose bribe amount B_i is greater than R_i. The smart contract consists of the function to confirm the receipt of message from those agents i and transfer the bribe amount B_i to the bribed agents. Each agent makes a net payment equal to his bribe value, minus the bribe amount received from all his descendants. Note that if an agent $i \in N$ could potentially receive the message from $c > 1$ number of agents possessing the message, then bribe value v_i of the agent i is equally distributed to all c agents by the smart contract. Similarly, if multiple agents do not receive bribe from a common descendant i, the reward value r_i is equally divided among the honest agents in our mechanism.

4.3 Example

Before analyzing the properties of our privacy mechanism SNPM, we study an example given in Fig. 3 to demonstrate our mechanism. Figure 3a shows a simple social network, where each node represents an agent, and m denotes the messenger who wishes to propagate the message to a subset of agents in the network. The edges between the nodes represent the neighbourhood relationship. The values provided alongside the nodes represent the valuations associated with each agent in the form of U_i/v_i, i.e., the first value represents the messenger valuation U_i and the value v_i represents the agent's valuation for the message. In this setting, we assume that all the agents truthfully report their valuations. Suppose the messenger wishes to propagate

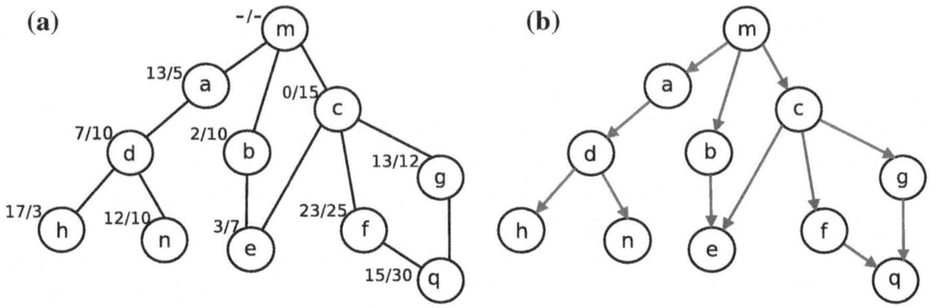

Fig. 3 **a** A social network example depicting the utilities of messenger/agents for a message forwarded by agent m. **b** The corresponding information diffusion network

a private message only to a subset of his neighbours in the network, i.e., agents b and c. The agent b has one descendant e and agent c has three descendants e, f and g who could potentially bribe b and c respectively for the private information. The recursive equation to obtain the cumulative bribe amount and reward for agents b and c is computed as follows:

$$R_b = U_b + \frac{1}{2} R_e.$$

$$B_b = v_b + \frac{1}{2} R_e.$$

$$R_c = U_c + R_f + (B_g + \epsilon) + \frac{1}{2} R_e.$$

$$B_c = v_c + R_f + (B_g + \epsilon) + \frac{1}{2} R_e.$$

We note that, as per our definition, both the agents b and c are potential sources of the private information for agent e, and hence the bribe value of e is equally distributed to both b and c. We recursively compute the same for the descendants e, f and g as follows:

$$R_f = U_f + \frac{1}{2} R_q.$$

$$B_f = v_f + \frac{1}{2} R_q.$$

$$R_g = U_g + \frac{1}{2} R_q.$$

$$B_g = v_g + \frac{1}{2} R_q.$$

The agent e have $R_e = U_e = 3$, $B_e = v_e = 7$ and $\pi_e = 1$. Similarly, agent q has $R_q = 15$, $B_q = 30$ and $\pi_q = 1$. Solving the recursive equations, we obtain $R_f = 30.5$, $B_f = 32.5$ and $\pi_f = 1$. Again, $R_g = 20.5$, $B_g = 19.5$ and $\pi_g = 0$. Finally, the net budget for agents b and c obtained is $R_b = 3.5$, $B_b = 11.5$ and $\pi_b = 1$, again $R_c = 51.5 + \epsilon$, $B_c = 66.5 + \epsilon$ and $\pi_c = 1$. Hence, agents e, f and q successfully receives the message by bribing via the smart contract as illustrated by the information diffusion network in Fig. 3b. In the figure, the green arrows denote the information diffusion flow, and the red arrows denote the forbidden flow to unintended agents. The budget of node g is not sufficient to bribe his neighbours c or q for the message, and hence he is forbidden from the message. As per the payment policy, the payment made by the agents e, q and f towards bribing their neighbours are $p_e = 7$, $p_q = 30$ and $p_f = 25$ respectively.

4.4 Properties

Theorem 1 The Social Network Privacy Mechanism is individually rational.

Proof Assume that an agent $i \in N_{-m}$ truthfully reports her bribe $v_i' \leq v_i$ to receive the message. If $v_i' > U_i$, agent i receives the message and her utility $u_i(a_i, a') > 0$, while if $U_i > v_i'$, the agent i does not receive the message and his payment due is zero according to the payment policy. If an agent i reports a bribe $v_i' > v_i$ such that $U_i < v_i'$, according to the allocation policy $\pi_i = 1$ and he receives the message. However his utility $u_i = (v_i - v_i') - p_i$ is negative. Therefore, for an arbitrary agent, when he truthfully reports her bribe, his utility is non-negative and SNPM is individually rational.

Theorem 2 The Social Network Privacy Mechanism is incentive compatible.

Proof To prove that SNPM is incentive compatible, we analyze the action of all the agents in the social network in the following two cases:

Case 1: If an agent $i \in N$ is forbidden from receiving the message, it indicates that $R_i > B_i$ and $\pi_i = 0$ according to the allocation policy. For any agent j who is a potential source of the message for node i, agent j receives an reward $B_i + \epsilon$ for not receiving a bribe from agent i, whereas it receives a lesser bribe value B_i from agent i. Therefore, agent j has no motivation in receiving a bribe from its descendant i as it does not maximize his utility. Hence, not propagating the message is a dominant strategy for an agent j whose descendant i exhibits $R_i > B_i$.

Case 2: If an agent $i \in N$ has sufficient budget to bribe an agent for receiving the message, it indicates that $R_i < B_i$ and $\pi_i = 1$ according to the allocation policy. For any agent j, who is a potential source for the message to node i, agent j receives an reward R_i for not receiving a bribe from agent i, whereas it receives a higher bribe value $B_i > R_i$ from agent i. Therefore, agent j has no motivation for stopping a message propagation to descendant i in return of a bribe value B_i, as it maximizes his utility. Hence, propagating the message is a dominant strategy for an agent j whose descendant i exhibits $R_i < B_i$ and is eligible to receive the message.

Therefore, for an agent $i \in N$ possessing the private message, propagating the message to only those agents who pay a higher bribe value than the reward offered and not propagating to the forbidden descendants is a dominant strategy. This ensures incentive compatibility for SNPM.

We note that the SNPM mechanism is *weakly budget balanced* which follows from the discussion in Sect. 4.1.

5 Conclusions and Future Work

In this paper, we generalized the mechanism design problem to the social network privacy setting, in which a sender sends a private message to a selected list of agents in the network, propagated to other agents through the neighbours of the sender. Our mechanism promotes message privacy by leveraging blockchain powered smart contracts to incentivize the receivers who do not disperse the message to their neighbours. Our privacy mechanism is novel in the sense that it employs simple game theoretic tools and distributed consensus mechanism to employ privacy in the social network, while satisfying all the necessary conditions for a mechanism to function. Our mechanism can function in a network involving multiple message propagation from multiple senders, as each message sharing is independent of the other, and does not create any conflict the system. The previous attempts to ensure privacy of user data in social networks were mainly achieved through centrally enforced policies and privacy systems that lacks trust and transparency, or employing a public ledger based distributed networking system to track private data, which suffers from scalability issues. Our approach positively resolves the problem of shared data privacy by employing simple albeit significant mechanism design principles.

We assume the underlying diffusion network to be a directed acyclic graph (DAG). The existence of algorithms for conversions of periodic structures to DAGs creates the possibility (although challenging) of an extension of our algorithm to generic graph structures. We leave it as an open problem. In addition to that, we plan to integrate time to our mechanism. That will decrease the possibility that our game may continue forever.

It would also be an interesting direction to efficiently incorporate mechanism design and blockchain technology in other aspects of privacy preservation in social networks, such as anonymization and privacy preservation of user information.

Acknowledgements Parts of this work is done while Arinjita Paul was conducting an internship supported by the JST Sakura Science Program. We would like to thank Prof. Satoru Iwata for hosting the internship. We also would like to thank anonymous reviewers who give comments that can significantly improve this paper.

References

1. Akasha: http://akasha.world/ (2015)
2. Azaria, A., Ekblaw, A., Vieira, T., Lippman, A.: Medrec: using blockchain for medical data access and permission management. In: 2nd International Conference on Open and Big Data, OBD 2016, Vienna, Austria, 22–24 August 2016, pp. 25–30 (2016)
3. Buterin, V., et al.: A next-generation smart contract and decentralized application platform. White paper (2014)
4. Casas-Roma, J., Rousseau, F.: Community-preserving generalization of social networks. In: Proceedings of the 2015 IEEE/ACM International Conference on Advances in Social Networks Analysis and Mining, ASONAM 2015, Paris, France, 25–28 August 2015, pp. 1465–1472 (2015)
5. Cheng, J., Fu, A.W.-C. Liu, J.: K-isomorphism: privacy preserving network publication against structural attacks. In: Proceedings of the 2010 ACM SIGMOD International Conference on Management of data, pp. 459–470. ACM, New York (2010)
6. Cohen, E., Megiddo, N.: Recognizing properties of periodic graphs. In: Applied Geometry and Discrete Mathematics, Proceedings of a DIMACS Workshop, Providence, Rhode Island, USA, 18 September 1990, vol. 4, pp. 135–146. DIMACS/AMS (1990)
7. Dagher, G.G., Mohler, J., Milojkovic, M., Marella, P.B.: Ancile: privacy-preserving framework for access control and interoperability of electronic health records using blockchain technology. Sustain. Cities Soc. **39**, 283–297 (2018)
8. Diaspora: https://diasporafoundation.org/ (2010)
9. Dwork, C.: The differential privacy frontier. In: Proceedings of Theory of Cryptography, 6th Theory of Cryptography Conference, TCC 2009, San Francisco, CA, USA, 15–17 March 2009, pp. 496–502 (2009)
10. Dwork, C., McSherry, F., Nissim, K., Smith, A.D.: Calibrating noise to sensitivity in private data analysis. In: Proceedings of Theory of Cryptography, Third Theory of Cryptography Conference, TCC 2006, New York, NY, USA, 4–7 March 2006, pp. 265–284 (2006)
11. Goel, V.: Facebook tinkers with users emotions in news feed experiment, stirring outcry. The New York Times (2014)
12. Gradwohl, R.: Information sharing and privacy in networks. In: Proceedings of the 2017 ACM Conference on Economics and Computation, EC '17, Cambridge, MA, USA, 26–30 June 2017, pp. 349–350 (2017)
13. Hardekopf, B.: The big data breaches of 2014. Forbes, 13 January 2015
14. Hay, M., Miklau, G., Jensen, D., Weis, P., Srivastava, S.: Anonymizing Social Networks. Computer Science Department Faculty Publication Series, p. 180 (2007)
15. Henson, B., Reyns, B.W., Fisher, B.S.: Security in the 21st century: examining the link between online social network activity, privacy, and interpersonal victimization. Crim. Justice Rev. **36**(3), 253–268 (2011)
16. Hongfei, D., Zhang, E.: https://docs.neo.org/en-us/whitepaper.html (2014)
17. Iwano, K., Steiglitz, K.: Testing for cycles in infinite graphs with periodic structure. In: Aho, A.V. (eds.) Proceedings of the 19th Annual ACM Symposium on Theory of Computing, pp. 46–55. ACM, New York (1987)
18. Kempe, D., Kleinberg, J.M., Kumar, A.: Connectivity and inference problems for temporal networks. In: Proceedings of the Thirty-Second Annual ACM Symposium on Theory of Computing, Portland, OR, USA, 21–23 May 2000, pp. 504–513. ACM, New York (2000)
19. Kleinberg, J.M., Raghavan, P.: Query incentive networks. In: Proceedings of 46th Annual IEEE Symposium on Foundations of Computer Science (FOCS 2005), 23–25 October 2005, Pittsburgh, PA, USA, pp. 132–141. IEEE Computer Society (2005)
20. Li, B., Hao, D., Zhao, D., Zhou, T.: Mechanism design in social networks. In: Proceedings of the Thirty-First AAAI Conference on Artificial Intelligence, San Francisco, California, USA, 4–9 February 2017, pp. 586–592 (2017)

21. Li, N., Li, T., Venkatasubramanian, S.: t-closeness: privacy beyond k-anonymity and l-diversity. In: IEEE 23rd International Conference on Data Engineering. ICDE 2007, pp. 106–115. IEEE (2007)
22. Luo, W., Liu, J., Liu, J., Fan, C.: An analysis of security in social networks. In: Eighth IEEE International Conference on Dependable, Autonomic and Secure Computing, pp. 648–651. IEEE (2009)
23. Machanavajjhala, A., Gehrke, J., Kifer, D., Venkitasubramaniam, M.: l-diversity: privacy beyond k-anonymity. In: Proceedings of the 22nd International Conference on Data Engineering. ICDE'06, pp. 24–24. IEEE (2006)
24. Moulin, H.: Axioms of Cooperative Decision Making. Econometric Society Monographs, vol. 15. Cambridge University Press, Cambridge (1991)
25. Nakamoto, S.: Bitcoin: a peer-to-peer electronic cash system (2008)
26. Nisan, N., Roughgarden, T., Tardos, E., Vazirani, V.V.: Algorithmic Game Theory, vol. 1. Cambridge University Press, Cambridge (2007)
27. Nissim, K., Smorodinsky, R., Tennenholtz, M.: Approximately optimal mechanism design via differential privacy. In: Innovations in Theoretical Computer Science, Cambridge, MA, USA, 8–10 January 2012, pp. 203–213 (2012)
28. Pai, M.M., Roth, A.: Privacy and mechanism design. SIGecom Exch. **12**(1), 8–29 (2013)
29. Rajagopalan, K.: Interacting with users in social networks: the follow-back problem. Technical report, MIT Lincoln Laboratory Lexington United States (2016)
30. Steemit social network: https://steemit.com/ (2016)
31. Sweeney, L.: k-anonymity: a model for protecting privacy. Int. J. Uncertain. Fuzziness Knowl.-Based Syst. **10**(05), 557–570 (2002)
32. Szabo, N.: Smart contracts (1994)
33. Vickrey, W.: Counterspeculation, auctions, and competitive sealed tenders. J. Financ. **16**(1), 8–37 (1961)
34. Williams, J.: Social networking applications in health care: threats to the privacy and security of health information. In: Proceedings of the 2010 ICSE Workshop on Software Engineering in Health Care, pp. 39–49. ACM, New York (2010)
35. Zyskind, G., Nathan, O., Pentland, A.: Decentralizing privacy: using blockchain to protect personal data. In: 2015 IEEE Symposium on Security and Privacy Workshops, SPW 2015, San Jose, CA, USA, 21–22 May 2015, pp. 180–184 (2015)

Balancing Cryptoassets and Gold: A Weighted-Risk-Contribution Index for the Alternative Asset Space

Aikaterini Koutsouri, Francesco Poli, Elise Alfieri, Michael Petch, Walter Distaso and William J. Knottenbelt

Abstract Bitcoin is foremost amongst the emerging asset class known as cryptoassets. Two noteworthy characteristics of the returns of non-stablecoin cryptoassets are their high volatility, which brings with it a high level of risk, and their high intraclass correlation, which limits the benefits that can be had by diversifying across multiple cryptoassets. Yet cryptoassets exhibit no correlation with gold, a highly-liquid yet scarce asset which has proved to function as a safe haven during crises affecting traditional financial systems. As exemplified by Shannon's Demon, a lack of correlation between assets opens the door to principled risk control through so-called volatility harvesting involving periodic rebalancing. In this paper we propose an index which combines a basket of five cryptoassets with an investment in gold in a way that aims to improve the risk profile of the resulting portfolio while preserving its independence from mainstream financial asset classes such as stocks, bonds and fiat currencies. We generalise the theory of Equal Risk Contribution to allow for weighting according to a desired level of contribution to volatility. We find a crypto–gold weighting based on Weighted Risk Contribution to be historically more effective in terms of Sharpe Ratio than several alternative asset allocation strategies including Shannon's Demon.

A. Koutsouri (✉) · W. J. Knottenbelt
Department of Computing, Imperial College London, London SW7 2AZ, UK
e-mail: k.koutsouri@imperial.ac.uk

W. J. Knottenbelt
e-mail: wjk@doc.ic.ac.uk

F. Poli · W. Distaso
Department of Finance, Imperial College London, London SW7 2AZ, UK
e-mail: f.poli@imperial.ac.uk

W. Distaso
e-mail: w.distaso@imperial.ac.uk

E. Alfieri
University Grenoble Alpes, 38000 Grenoble, France
e-mail: Elise.Alfieri@univ-grenoble-alpes.fr

M. Petch
CoinShares, Octagon Point, 5 Cheapside, St Paul's, London EC2V 6AA, England
e-mail: mpetch@coinshares.co.uk

P. Pardalos et al. (eds.), *Mathematical Research for Blockchain Economy*,
Springer Proceedings in Business and Economics,
https://doi.org/10.1007/978-3-030-37110-4_15

Within the crypto-basket, whose constituents are selected and rebalanced monthly, we find an Equal Weighting scheme to be more effective in terms of the same metric than a market capitalisation weighting.

Keywords Cryptoassets · Index · Volatility · Rebalancing premium · Risk management · Asset allocation · Equal risk contribution · Weighted risk contribution · Bitcoin · Gold

1 Introduction

Cryptoassets are increasingly recognised as viable investments. Since they exhibit little correlation with traditional asset classes [1], adding even a small percentage of cryptoassets to a portfolio can enhance risk-adjusted returns [6] while also offering a safe haven in the event of a financial crisis. Until recently, the cryptoasset market was afflicted by a lack of clear regulatory guidance, with uncertainty surrounding the classification of the asset, its tax treatment, and the effect of events specific to the cryptocurrency domain such as forks. However, the last year has seen the emergence of more mature and well-defined regulatory frameworks [11, 17, 20]. In turn, this has driven increased institutional demand for cryptoasset-based financial indicators.

Regardless of the creation of new financial products, many investors still see the crypto market as being unacceptably risky due to its high volatility—something not unusual for an emerging asset class. Although volatility poses challenges in terms of increased uncertainty, there are also benefits to be had from its proper management through diversification and regular rebalancing [4]. This is exemplified by the so-called Shannon's Demon approach in which two, ideally uncorrelated, assets—at least one of which is highly volatile— are periodically rebalanced to maintain an ideal target allocation. The resulting expected growth rate is greater than the arithmetic mean of the individual expected growth rates, while the variance of the returns is less than the mean of the individual variances [15, pp. 201–209].

In theory, this strategy would be well-suited for the volatile cryptoasset class and an uncorrelated wealth-preserving asset class. Although there are plenty of candidates uncorrelated with cryptoassets, not all are properly suited. For example, traditional wealth-preserving assets such as property or museum-quality fine art are illiquid [16]. An asset such as gold is much more appropriate because of its low volatility, high liquidity and ability to act as a hedge to traditional financial markets [1, 5, 10]. Gold is also more suitable in this context than other precious metals such as platinum or silver, since the latter, unlike gold, have not historically served as a hedge or safe haven during times of financial turmoil [12].

Pure-crypto indices such as CRIX [21], CRYPTO20 [7], MVDA5 [14], and Bloomberg Galaxy Crypto Index [2] do offer broad exposure to the crypto-market. However, they are characterised by a volatility close to that of the single cryptoassets. Thus, they do not incorporate mechanisms for effective risk control beyond simple diversification over their (highly-correlated) constituents.

By contrast, the purpose of this study, put forth jointly by researchers at Imperial College London and CoinShares, is to propose a low-volatility index that combines an uncorrelated asset (gold) with a basket of cryptoassets, using weighted-risk contribution as a rebalancing mechanism. By decreasing volatility levels, it yields superior risk-adjusted returns when compared to a number of alternative strategies, including holding cryptoassets or gold alone. Further, the proposed index presents a moderate turnover, which translates into moderate operating costs. A fuller index methodology document, together with a reference implementation, will be made available online in due course.

The remainder of this paper is organised as follows. Section 2 presents background related to portfolio diversification, while Sect. 3 extends the theory of Equal Risk Contribution to Weighted Risk Contribution, in which the contribution to volatility by two uncorrelated asset classes (cryptoassets and gold in our case) can be varied to taste. Section 4 presents an overview of the index methodology. Section 5 demonstrates the historical performance of the index, presenting an improved risk and returns profile compared to other established methodologies. Section 6 concludes.

2 Background

2.1 Shannon's Demon

In the 1960s, Claude Shannon presented an optimal growth-portfolio construction method exploiting diversification and rebalancing (cf. [3, 19]). This method, called Shannon's Demon [15], considers two assets: a highly volatile one that follows a pure random walk process, which can either double in price or drop by 50%, and an uncorrelated low-volatility one, specifically cash. In the proposed portfolio half of the capital is allocated to the volatile asset and half to cash.

The general setting is: an investor decides at the beginning which fraction of his initial wealth to put at risk and then regularly rebalances between the risky asset and cash taking into account the proceeds in the game. The accumulated wealth after T rounds and L losses is $W_T = W_0[(1 + w_0 a)^{(T-L)}(1 - w_0 b)^L]$, where W_0 is the initial wealth, w_0 the fraction reallocated in risk, a the percentage returns in an up-move proportional gain and b the percentage loss in a down-move proportional loss. The objective is to maximise the log utility of wealth, which implies maximising the expected growth rate:

$$E[g] = p \log(1 + w_0 a) + q \log(1 - w_0 b)$$

where p is the probability of profit and q is the probability of loss. The optimal fraction to invest in the risky asset is then:

$$w_0^* = \frac{pa - qb}{ab}$$

The conditions under which a rebalancing approach outperforms buy-and-hold are explored in [9]. The same work considers a case of two negatively-correlated volatile assets. It is shown that even with diversification and positive expected returns, a buy-and-hold strategy can fail to grow an investor's wealth. Active management by rebalancing, on the other hand, builds long-term value.

Shannon's scheme is a strategy that can generate growth even if the returns of both assets are negative. It provides a solution to Parrondo's Paradox [18] which states that a winning strategy can emerge from the intelligent combination of two losing strategies. More generally, it provides a means to harness high volatility and low correlation in a way that reduces portfolio risk through diversification and rebalancing [4]. However, there is a trade-off between the frequency of rebalancing (which should be high to lead to a higher growth rate) and turnover (which should be kept moderate to avoid high transaction costs).

2.2 Equal Risk Contribution

Risk-based strategies have proved to be capable of reducing volatility in a way that does not impede market exposure, while outperforming standard strategies in unsteady markets. Equal Risk Contribution (also known as Risk Parity) is a well known risk-control strategy that achieves diversification both within and across asset classes. Its main goal is to bolster the portfolio's immunity to unforeseen drawdowns during stressful market periods. In contrast with the equally-weighted allocation scheme, a Risk Parity portfolio aims towards an equal distribution of the overall budget, expressed in terms of risk rather than capital. As a result, it achieves better risk-adjusted returns.

The Risk Parity optimisation problem setting is constituted of $N \geq 2$ assets $A_1, \ldots, A_N$, with μ_i, σ_i and σ_i^2 representing the expected return, standard deviation and variance of the returns of A_i respectively and ρ_{ij} denoting the correlation coefficient of the returns of A_i and A_j for $i \neq j$. The $N \times N$ symmetric covariance matrix of returns is defined as $\Sigma = (\sigma_{ij})$ where $\sigma_{ij} = \rho_{ij}\sigma_i\sigma_j, i \neq j$ and $\sigma_{ij} = \sigma_i^2, i = j$. If x_i is the amount to be invested in asset A_i, then the volatility (measured in terms of standard deviation) of the resulting portfolio $x = (x_1, \ldots, x_i)$ is computed as $\sqrt{x^T \Sigma x}$.

In the Equal Risk Contribution problem, $\sigma(x) = \sqrt{x^T \Sigma x}$ denotes the risk of the portfolio, and through the Euler decomposition, the risk is expressed as:

$$\sigma(x) = \sum_{i=1}^{N} \sigma_i(x) = \sum_{i=1}^{N} x_i \times \frac{\partial \sigma(x)}{\partial x_i},$$

where $\partial_{x_i}\sigma(x)$ is the marginal risk contribution and $\sigma_i(x) = x_i \times \partial_{x_i}\sigma(x)$ is the (total) risk contribution of asset A_i. The desired risk-balanced portfolio is constituted in a way that all components contribute equally to the overall volatility; therefore $\sigma_i(x) = \sigma_j(x)$. The general Risk Parity portfolio construction problem can be mathematically expressed as:

$$x_{ERC} = \left\{ x \in [0,1]^N : x_i \times \partial_{x_i}\sigma(x) = x_j \times \partial_{x_j}\sigma(x), \forall i, j, \sum_{i=1}^{N} x_i = 1 \right\}.$$

Through the problem expression, asset classes with reduced levels of volatility or correlation are favoured since their marginal risk contribution to the portfolio volatility will be lower. In [13], Maillard et al. show that if all correlations are the same then each constituent weight is defined as the ratio of the reciprocal of its volatility with the sum of the reciprocals of the volatilities of all constituents:

$$x_i = \frac{\sigma_i^{-1}}{\sum_{j=1}^{N} \sigma_j^{-1}}, i = 1, \ldots, N \tag{1}$$

and therefore, in the bivariate case,

$$x_1 = \frac{\sigma_1^{-1}}{\sigma_1^{-1} + \sigma_2^{-1}}.$$

In [13], when the correlations are different, the authors propose solving the optimisation problem defined as

$$\min_x \sum_{i=1}^{N} \sum_{j=1}^{N} \left(x_i (\Sigma x)_i - x_j (\Sigma x)_j \right)^2 \tag{2}$$

with $x_i \in [0, 1]$ and $\sum_{i=1}^{N} x_i = 1$. Here $(\Sigma x)_i$ denotes the ith entry of the vector resulting from the product of Σ with x.

3 Weighted Risk Contribution

One potential concern about the classic Equal Risk Contribution scheme is that, because it belongs to the family of inverse volatility weighting, it can potentially generate allocations that are too concentrated towards assets with low volatility or low correlation, causing the undesired effect of lowering the degree of diversification inside a portfolio when no constraints are introduced.

This is indeed what happened to many Risk Parity funds in the last two years. In fact, given the low rates set by Central Banks in the most advanced economies, sovereign bonds returns reached an unprecedented low level of volatility and an unconstrained minimisation resulted in an extremely high weight for this asset class. When Central Banks moved on to raise rates, Risk Parity portfolios found themselves too exposed to that risk and suffered important losses. In this case, when it comes to the weighting of cryptoassets alone, the risk of a similar scenario is somehow less of a concern, because of the similar level of volatility between cryptoassets and because of their high level of correlation.

We address this issue by allowing the proportion of risk contribution by each asset class to be configurable. Following [13], the vector of risk contributions in the two-asset case given weighting $x = (x_1, x_2)$ and correlation ρ is:

$$\frac{1}{\sigma(x)} \begin{pmatrix} x_1^2\sigma_1^2 + x_1x_2\rho\sigma_1\sigma_2 \\ x_2^2\sigma_2^2 + x_1x_2\rho\sigma_1\sigma_2 \end{pmatrix}$$

Considering the case of uncorrelated assets ($\rho = 0$), and supposing that we desire the risk contribution of asset 1 to be α times the risk contribution of asset 2, we need to solve for x_1 in:

$$x_1^2\sigma_1^2 = \alpha\left(x_2^2\sigma_2^2\right)$$

Given $x_i \in [0, 1]$ and $\sum_{i=1}^{2} x_i = 1$ this yields:

$$x_1 = \frac{\sqrt{\alpha}\,\sigma_1^{-1}}{\sqrt{\alpha}\,\sigma_1^{-1} + \sigma_2^{-1}} \tag{3}$$

In our case, x_1 represents the proportion of the investment allocated towards a basket of cryptoassets whose components are equally weighted, while x_2 is the proportion invested in gold. The risk contribution ratio is set as $\alpha = 4$, indicating that 80% of the total risk emanates from the crypto-basket.

4 Index Overview

4.1 Design Goals

The objective of this study is the design and implementation of an index that should meet the following goals:

1. Provide exposure to the alternative asset space in a way that is orthogonal to traditional financial markets;
2. Be comprised of a small number of liquid, investable constituent assets;
3. Exhibit a relatively stable composition in terms of constituents with asset weights that do not vary dramatically between rebalancing periods, leading to low or moderate turnover;

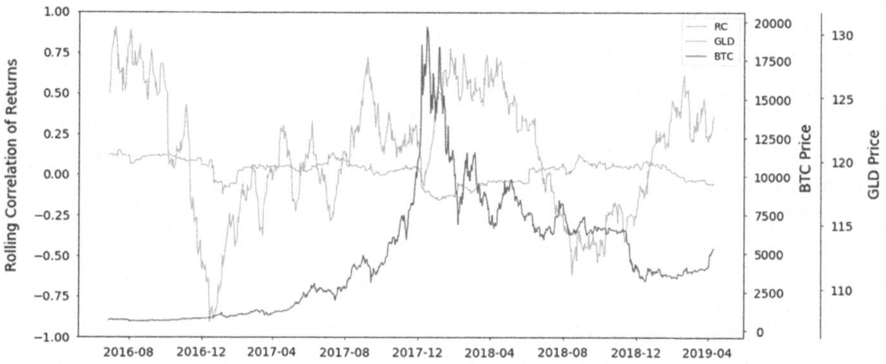

Fig. 1 180-day rolling correlation (RC) between daily returns of Bitcoin (BTC) and Gold (GLD)

4. Utilise some means of principled risk control leading to lower volatility;
5. Be specified in a clear and unambiguous manner to facilitate validation and reproducibility;
6. Hold constituent assets on a long-only basis;
7. Not make use of leverage.

In terms of Goal 4, historical volatility of cryptoassets has remained at much higher levels compared to other asset classes while correlation among single non-stablecoin cryptoassets is persistent, displaying some signs of time variability. Therefore, constructing an index constituted only of cryptoassets offers very little prospect of diversification irrespective of the methodology used and hence, less prospect of bringing down its volatility. Gold returns, on the other hand, have been much less volatile than those of cryptoassets and have displayed a very low time varying correlation with cryptoassets (see Fig. 1). Gold was therefore the ideal candidate to include alongside cryptoassets with the purpose of considerably reducing volatility.

4.2 Constituent Eligibility and Selection

The index is composed of a fixed number of constituents including five cryptoassets and SPDR Gold Shares (GLD), the largest gold ETF. The cryptoasset constituents of the index are the top five eligible cryptoassets based on the 6-month rolling mean of free-float market capitalisation. By restricting the index to the top five cryptoassets we are less likely to encounter liquidity issues. Selection of constituents occurs on a monthly basis.

We determine whether a cryptoasset is eligible to be selected, based on the following requirements:

1. Trades in USD;
2. Is not linked to the value of a fiat currency;

3. Has at least a 6-month history of trading on a reputable exchange;
4. Has been on its native blockchain for at least 6 months;
5. Is not an ERC20 token;
6. Is not a privacy-focused coin (e.g. Monero, ZCash);
7. Has not suffered a major chain reorganisation in the last 6 months, and is not subject to a forthcoming contentious hard fork before the next selection is due to take place.

4.3 Constituent Weighting

For the weighting of the constituents, we choose a bi-level approach that involves studying the historical volatilities of the crypto-basket and gold separately in order to inform the crypto–gold asset allocation decision. That is because if GLD is added to a basket of five cryptoassets for a global allocation scheme, the correlation structure between all six assets cannot be ignored and the constituents' weighting procedure cannot be performed through Eq. (1). Also, in order to be able to produce a robust estimation of covariance matrices, the behaviour of the two asset classes would have to be studied only in time spans where exchanges for both are open. The bi-level approach on the other hand allows for exploitation of all available market data.

Regarding the formation of the crypto-basket, due to the persistent levels of correlation between non-stablecoin cryptoassets, any Risk Parity approach is expected to lean towards an Equally Weighted allocation whose risk level is not significantly improved. Therefore, due to its much more convenient reproducibility compared to Eq. (2) and the fragility of Eq. (2) when the covariance matrix is barely positive semi-definite, an Equally Weighted scheme is employed within the crypto-basket.

Taking into consideration the former, and the lack of a significant correlation between gold and cryptoassets, the index is calculated following a two-stage allocation scheme that involves:

1. Computation of the historical volatility of (a) the equally weighted crypto-basket, and (b) gold;
2. Asset allocation among the crypto-basket and gold expressed as the bivariate weighted risk contribution problem presented in Sect. 3.

4.4 Rebalancing Schedule

In order to capture the diversification benefits of the time-varying correlations between gold and crypto highlighted in Fig. 1, we have chosen a monthly rebalancing frequency. Coupled with the monthly reselection it allows the index to represent the rapidly evolving market conditions. This has no dramatic impact on the turnover of the portfolio and hence keeps transaction costs to an acceptable level.

4.5 Index Calculation

The Index base level is set on 1 000 on January 1st, 2016:

$$\text{Index}_0 = 1\,000 \tag{4}$$

The Index level on day t from January 2nd, 2016 onwards is calculated as:

$$\text{Index}_t = \frac{\sum_{i \in N_t} P_{i,t} \times x_{i,t}}{D_t} \tag{5}$$

where

- N_t is the set of the 6 selected assets (5 cryptocurrencies and gold) on day t
- $P_{i,t}$ is the closing price for asset i on day t expressed in USD
- $x_{i,t}$ is the weight of asset i on day t as computed through the WRC allocation scheme at the beginning of the month
- D_t is the Index Divisor on day t.

The Index Divisor is used so that assets weight rebalancing and substitution do not alter the Index level. It is calculated using the following formula:

$$D_t = \frac{\sum_{i \in N_t} P_{i,t-1} \times x_{i,t}}{\sum_{i \in N_{t-1}} P_{i,t-1} \times x_{i,t-1}} \times D_{t-1} \tag{6}$$

The Divisor on January 2nd, 2016 is calculated as:

$$D_1 = \frac{\sum_{i \in N_1} P_{i,0} \times x_{i,1}}{1\,000} \tag{7}$$

where

- N_1 is the set of the selected assets on January 2nd, 2016
- $P_{i,0}$ is the closing price for asset i on January 1st, 2016 expressed in USD
- $x_{i,1}$ is the weight of asset i on January 2nd, 2016.

Equations (5–7) are equivalent to computing recursively the value of the Index using the weighted average of its constituent's returns:

$$\text{Index}_t = \sum_{i \in N_t} \frac{P_{i,t}}{P_{i,t-1}} x_{i,t} \times \sum_{i \in N_{t-1}} \frac{P_{i,t-1}}{P_{i,t-2}} x_{i,t-1} \times \cdots \times \sum_{i \in N_1} \frac{P_{i,1}}{P_{i,0}} x_{i,1} \times 1\,000 \tag{8}$$

which implies

$$\text{Index}_t = \sum_{i \in N_t} \left(1 + R_{i,t}\right) x_{i,t} \times \text{Index}_{t-1}, \quad t = 1, 2, \ldots \tag{9}$$

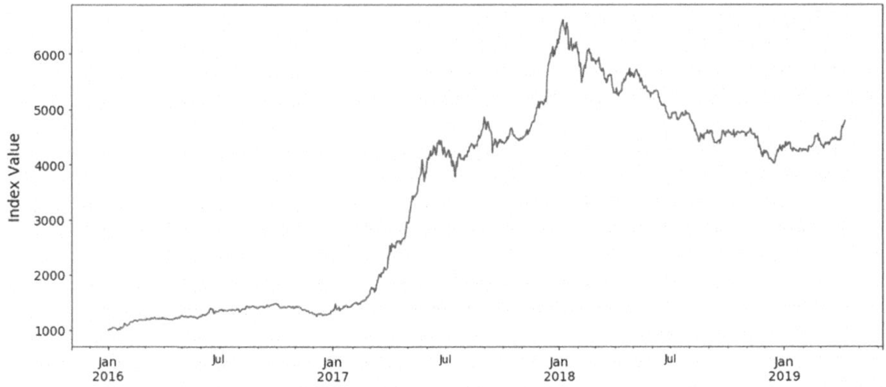

Fig. 2 Index value January 2016–April 2019

where

- $R_{i,t}$ is the return of asset i from time $t-1$ to time t
- Index_0 is the base level of the Index set at 1 000 on January 1st, 2016.

Figure 2 shows how the index value would have evolved over the period January 2016 to April 2019. A detailed breakdown and comparison of index performance is presented in Sect. 5.

4.6 Hard Fork and Airdrop Policy

Hard Fork Policy A 'Hard Fork' occurs when a change is made to the transaction validation rules of a cryptoasset's underlying blockchain protocol in a way that is not compatible with its earlier version. Nodes that wish to continue to participate are expected to upgrade to the new version of the protocol's software. Usually such a fork is planned and accepted by the overwhelming majority of nodes. However, where the fork is contentious enough that a non-negligible number of nodes continue to run the old version of the software, a chain split occurs.

The index will feature a Governing Committee which will evaluate all upcoming hard forks, especially in light of Rule 7 of Sect. 4.2. Treatment of hard forks will be led by decisions of exchanges with respect to the ticker symbols used to represent the resulting cryptoassets and the markets that they maintain. Concretely, suppose some cryptoasset traded under ticker symbol T is expected to undergo (or undergoes) a hard fork resulting in an original chain C with cryptoasset C_a and a modified chain C' with cryptoasset C'_a. There are a few scenarios to consider:

- C_a continues to trade under ticker symbol T while C'_a starts trading under a newly-created ticker symbol T'. The BTC–BCH fork is an example of this scenario. In this case, C_a continues as a constituent of the index. C'_a is not eligible to become a constituent of the index (lacking as it does the necessary pricing history), and does not contribute to the index value. C_a' may be sold by funds tracking the index as an excess return; the precise decision of when (or whether) to sell will be a matter of judgment for the tracking funds.

- C_a' now trades under ticker symbol T while C_a starts trading under a new ticker symbol T'. The ETH–ETC fork is an example of this scenario. In this case, C_a' replaces C_a as a constituent of the index. The pricing history for C_a' is taken as being that of C_a prior to the fork. C_a is no longer a constituent of the index, does not contribute to the index value, and may be sold by funds tracking the index as an excess return.
- C_a' now trades under ticker symbol T while trading in C_a is (largely) abandoned. Hard forks to upgrade the consensus mechanism of Monero usually follow this pattern. In this case, C_a' replaces C_a as a constituent of the index and the pricing history for C_a' is taken as being that of C_a prior to the fork.
- There is substantial disagreement amongst exchanges as to the ticker symbols that C_a' and C_a should trade under. Usually this scenario would arise as the result of a contentious hard fork. Since cryptoassets due to undergo contentious hard forks before the next selection date are not eligible for selection, it is expected that this situation would apply to index constituents only in very rare circumstances. In this case, an extraordinary meeting of the Governing Committee will be convened in order to decide on an appropriate course of action which may include replacing C_a by the next eligible cryptoasset, or rebalancing across the remaining constituent cryptoassets.

Airdrop Policy An Airdrop occurs when a blockchain project distributes free cryptoassets to investors in the hopes of attracting more people to use their platform. Occasionally some projects offer more established cryptoassets to do an Airdrop but most of the time, the project Airdrops their own native token or cryptocurrency. Requirements to qualify for an Airdrop vary as well; in some cases the participant has to hold the cryptoasset in their wallet while other times they have to promote the project on an online forum.

Airdropped cryptoassets will not be included in the index. Fund managers tracking the index may sell these at their earliest convenience, thus contributing to excess returns over the base index.

5 Results

5.1 Methodology and Data Source

In order to evaluate the effectiveness of a Weighted Risk Contribution (WRC) strategy in the cryptoasset and gold case, the performance of a respective risk distribution portfolio is measured and compared against various strategies including buy-and-hold bitcoin (BTC), buy-and-hold gold (GLD), market capitalisation weighted pure

Table 1 Top 5 eligible cryptoassets—monthly reselection

Date	Constituent 1	Constituent 2	Constituent 3	Constituent 4	Constituent 5
2016-01-01	Bitcoin (BTC)	Ripple (XRP)	Litecoine (LTC)	Dash (DASH)	Dogecoin (DOGE)
2016-03-01	BTC	XRP	LTC	Ethereum (ETH)	DASH
2017-02-01	BTC	ETH	XRP	LTC	Ethereum Classic (ETC)
2017-04-01	BTC	ETH	XRP	LTC	DASH
2017-07-01	BTC	ETH	XRP	LTC	ETC
2017-09-01	BTC	ETH	XRP	LTC	DASH
2018-03-01	BTC	ETH	XRP	Bitcoin cash (BCH)	LTC
2018-11-01	BTC	ETH	XRP	Stellar (XLM)	LTC
2019-01-01	BTC	ETH	XRP	BCH	EOS (EOS)

cryptoassets, Shannon's Demon using bitcoin and gold, and Equal Risk Contribution (ERC) cryptoassets. The dataset used for the implementation and backtesting of the described allocation method includes daily values of historical free float market capitalisation and USD prices for more than 3 000 cryptoassets, obtained from CoinGecko as well as daily adjusted USD prices of SPDR Gold Shares (GLD). The backtest that is performed covers the period between January 2016 and April 2019, a time span that reflects a wide variety of market conditions for the cryptoasset space. The datasets produce daily returns for both asset classes and assumes monthly rebalancing for all active strategies.

Table 1 shows the results of monthly selection of cryptoasset constituents that meet the eligibility criteria. Note that only dates where the constituents change are presented.

The crypto-basket composition is defined according to an Equally Weighted scheme, whose historical returns and volatility are studied towards the dynamic allocation between the cryptoassets and gold. We opt for a WRC allocation scheme between the two classes. Given the historical level of correlation between gold and crypto assets, an equal risk distribution among the two asset classes would be expected to be heavily concentrated towards gold as the lower volatility asset. Nevertheless, the chosen WRC setting, with a risk ratio that results to 80% of the total risk emanating from the crypto-basket component ($\alpha = 4$), ensures a good level of diversification, balancing the two components in the denominator of Eq. (3) as seen in Fig. 3.

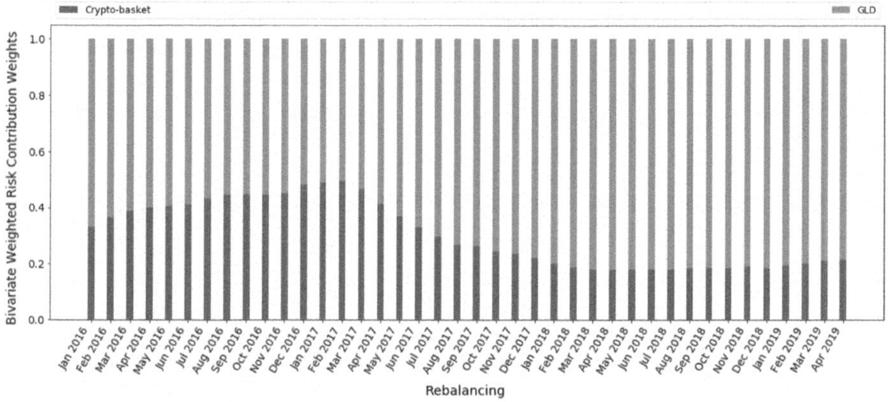

Fig. 3 Weighted risk contribution allocation—EW crypto basket base—$\alpha = 4$

5.2 Analysis and Results

The results obtained from the bivariate WRC allocation with an EW crypto-basket base (WRC-EW Base) are directly compared with the following:

1. Bivariate WRC allocation with a 6-month rolling mean Market Capitalisation weighted crypto-basket base (WRC-MC Base)
2. Equally-weighted cryptoassets (EW)
3. Market capitalisation weighted cryptoassets (MC)
4. Equal Risk Contribution weighted cryptoassets (ERC)
5. Bitcoin and GLD weighted in accordance with the Shannon's Demon (SD)
6. Bitcoin only (BTC)
7. Gold only (GLD).

As seen in Table 2 and Fig. 4, the proposed allocation scheme outperforms the rest in terms of historical risk-adjusted returns, as measured by the Sharpe Ratio. Moreover, a comparison with a typical index profile of the cryptoasset space, namely the MVIS Digital Assets 5 Index [14] (MVDA5)—a market capitalisation weighted index which tracks the performance of the five largest and most liquid cryptoassets—also reveals superiority in terms of the risk–return profile. Annualised returns are higher than a buy-and-hold GLD-only investment while annualised volatility levels are much lower than the crypto-market's. The ERC and EW present similar behaviour due to the assets' correlation structure; similarly, passive bitcoin and Market-Cap driven strategies do not reveal major differences. Table 2 also reports portfolio turnover, which reflects the total proportion of portfolio value traded (bought and sold) while rebalancing the portfolio, on an annualised basis as defined in [8].

Overall, the bivariate WRC allocation's performance is characterised by significantly lower volatility, and a more stable risk profile. The stability of the strategy's performance is further reflected in Fig. 5.

Table 2 Annualised performance of allocation schemes, Jan 2016–Apr 2019

Allocation scheme	Annualised returns	Annualised volatility	Annualised sharpe ratio	Annualised turnover
WRC-EW base	0.4797	0.2411	1.9894	1.6906
WRC-MC base	0.3199	0.2318	1.3800	1.1196
ERC	0.9878	0.8062	1.2253	2.8512
EW	1.0284	0.8292	1.2252	2.9556
MC	0.6734	0.7560	0.8908	1.2240
SD	0.4085	0.3889	1.0502	1.0800
BTC	0.7629	0.7680	0.9934	0.0000
GLD	0.0812	0.1452	0.5591	0.0000
MVDA5	1.1160	0.8757	1.2744	Not computed

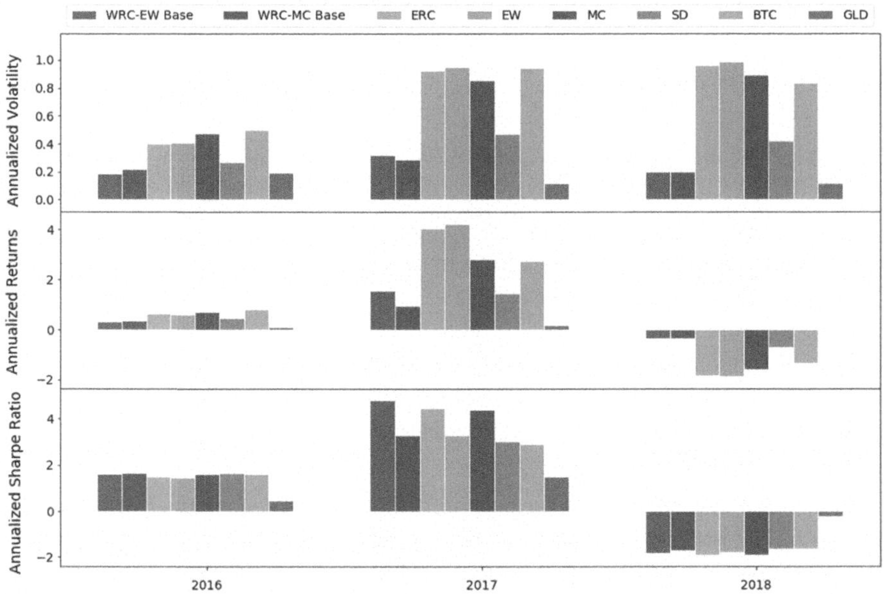

Fig. 4 Annualised returns, volatility and sharpe ratio Jan 2016–Dec 2018

6 Conclusion

We have proposed the construction of an index that offers investors exposure to alternative assets. By exploiting the characteristics of the two asset classes of cryptoassets and gold—namely the extremely high volatility of the former, the low volatility of the latter and the lack of correlation between the two—it is characterised by an attractive ability to reduce price instability while raising the average return per unit of volatility. By generalising the theory of equal risk contribution, we offer a sophisticated,

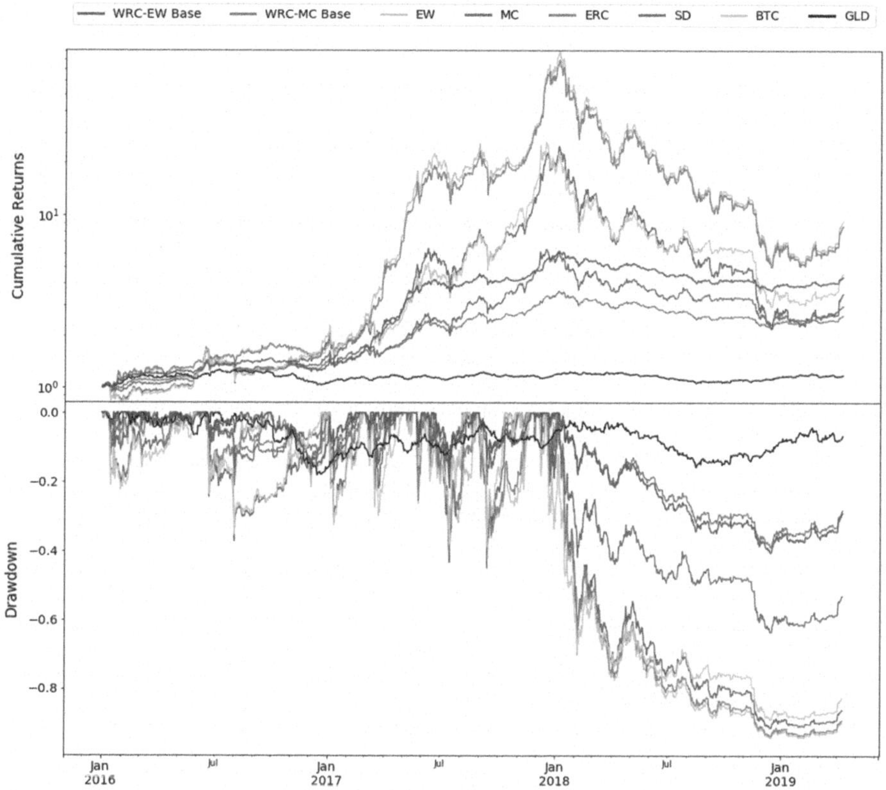

Fig. 5 Cumulative returns and drawdown Jan 2016–Apr 2019

albeit intuitive, way of tuning the exposure of an index to uncorrelated asset classes. Another important feature of the index lies in the associated moderate turnover, which translates into moderate operating costs. Finally, by taking into account a variety of events unique to the cryptoasset space such as hard forks and airdrops and by proposing corresponding policies, we have designed an investable product whose distinctive elements make it a unique form of investment.

Acknowledgements Imperial College of London gratefully acknowledges the support given mutually by CoinShares and Elwood Capital Management. Their funding provides Imperial with the support needed to conduct the research found within this paper.

References

1. Baur, D.G., Dimpfl, T., Kuck, K.: Bitcoin, gold and the US dollar—a replication and extension. Financ. Res. Lett. **25**, 103–110 (2018)
2. Bloomberg: Index methodology: bloomberg galaxy crypto index. Index Methodology (2019)
3. Booth, D., Fama, E.: Diversification returns and asset contributions. Financ. Anal. J. **48**(3), 26–32 (1992)
4. Bouchey, P., Nemtchinov, V., Paulsen, A., Stein, D.: Volatility harvesting: why does diversifying and rebalancing create portfolio growth? J. Wealth Manag. **15**(2), 26–35 (2012)
5. Bredin, D., Conlon, T., Poti, V.: Does gold glitter in the long-run? Gold as a hedge and safe haven across time and investment horizon. Int. Rev. Financ. Anal. **41**, 320–328 (2015)
6. Burniske, C., Tatar, J.: Cryptoassets: The Innovative Investor's Guide to Bitcoin and Beyond, 1st edn. McGraw-Hill Education, New York (2017)
7. CRYPTO20: the first tokenized cryptocurrency index fund. White Paper, October 2017
8. Demiguel, V., Garlappi, L., Uppal, R.: Optimal versus naive diversification: how inefficient is the 1/n portfolio strategy? Rev. Financ. Stud. **22**(05) (2009)
9. Dubikovsky, V., Susinno, G.: Demystifying rebalancing premium and extending portfolio theory in the process (2015). https://doi.org/10.2139/ssrn.2927791
10. Dyhrberg, A.: Hedging capabilities of bitcoin. Is it the virtual gold? Financ. Res. Lett. **16**(C), 139–144 (2016)
11. Treasury, H.M.: Financial conduct authority, Bank of England. Cryptoassets Task Force: Final Report, October (2018)
12. Hood, M., Malik, F.: Is gold the best hedge and a safe haven under changing stock market volatility? Rev. Financ. Econ. **22**(2), 47–52 (2013)
13. Maillard, S., Roncalli, T., Teiletche, J.: On the properties of equally-weighted risk contributions portfolios (2008). https://doi.org/10.2139/ssrn.1271972
14. MV Index Solutions: MVIS CryptoCompare Digital Assets 5 Index. Index Factsheet (March 2019)
15. Poundstone, W.: Fortune's Formula. Hill and Wang (2005)
16. Rickards, J.: The New Case for Gold. Portfolio/Penguin Random House, UK (2016)
17. Security and exchange Commission: virtual currencies: the oversight role of the U.S. securities and exchange Commission and the U.S. Commodity futures trading Commission. Report, February (2018)
18. Shu, J.J., Wang, Q.W.: Beyond Parrondo's paradox. Scientific Reports **4**, 4244 (2014)
19. Stein, D., Nemtchinov, V., Pittman, S.: Diversifying and rebalancing emerging market countries. J. Wealth Manag. **11**(4), 79–88 (2009)
20. The law Library of Congress: regulation of cryptocurrency in selected jurisdictions. Report, June (2018)
21. Trimborn, S., Härdle, W.K.: CRIX an index for cryptocurrencies. J. Empir. Financ. **49**, 107–122 (2018)

An Introduction to the Use of zk-SNARKs in Blockchains

Alexandre Miranda Pinto

Abstract The advent of blockchain brings a wide horizon of opportunities to the world. The first such example, Bitcoin, strongly advocates a principle of total openness, which is reflected in the fact that all transactions are public and the history of each account can easily be reconstructed. Although an account cannot immediately be linked to a real-world identity, this does not grant strong guarantees of anonymity, and such a feature of Bitcoin and similar blockchains prevents it from reaching wide acceptance for financial use-cases, where users often desire strong confidentiality of their balances and financial history. As a consequence, there has recently been a growing interest in privacy-enhancing technologies that ensure public permissionless blockchains can keep the details of transactions private according to particular use cases. One of the most promising technologies in this area is Zero-Knowledge Proofs, and in particular zk-SNARKs, due to their very short proofs and verification times. This makes them well suited to be used as transaction data, hiding all the private details at the same time they guarantee the integrity and accuracy of the transaction, and to be verified on-chain by a smart contract. This paper is an introductory presentation of this topic, what advantages zk-SNARKs bring to the blockchain ecosystem and how they can be tailored to specific applications.

Keywords Zero-knowledge proofs · zk-SNARKs · Privacy in blockchains

1 Introduction and Paper Layout

Since the introduction of Bitcoin in 2009 [28], Distributed-Ledger Technology (DLT), more often known as blockchain, has steadily grown and been recognized as a new tool with potentially revolutionary use-cases. Some of them will be mainly technical, and will harness the strong guarantees of distributed consensus and the maintenance of a single source of truth shared by many independent (collaborative,

A. M. Pinto (✉)
Artos Systems, 160 Fleet Street, London EC4A 2DQ, UK
e-mail: alex.miranda.pinto@gmail.com

P. Pardalos et al. (eds.), *Mathematical Research for Blockchain Economy*,
Springer Proceedings in Business and Economics,
https://doi.org/10.1007/978-3-030-37110-4_16

or possibly competing) parties; but others may be rooted on innovative ways of thinking social and economical relationships, such as the intrinsic value of money, whom we must trust to manage its creation and whether users should be allowed to transfer it outside any controls of central authorities.

One of the radical innovations of Bitcoin was the total openness of its ledger. There may be different reasons for this, for example a desire to remove all trust necessary in central authorities, who may be viewed as potentially corruptible entities bent on limiting individual freedoms; or instead a belief in total transparency as a virtue of advanced societies, where those who have nothing to hide should not fear scrutiny; or it may even have been a simple decision to solve a hard technical problem, Byzantine Agreement (see [25]), in a network with millions of participants.

Whatever the case, such absolute transparency is not always desired, especially if Bitcoin is intended as a replacement for fiat currencies as a means of exchange. Typical users value their privacy in this domain, and would rather prefer to keep their transactions history private. This is why it is commonly advised to use each Bitcoin address only once,[1] to avoid linking transactions to one same identity. A better way is to use privacy-enhanced blockchains, either where confidentiality has been designed in by default (for example ZCash,[2] Monero,[3] Grin,[4] Beam,[5] Dash[6]) or when it has been added a posteriori by some other mechanism. See for example Zether [11], Mimblewimble [24, 30] or Coinjoin [17].

Zero-Knowledge Proofs (ZKP) are among the most popular technologies, and have turned from a quite specialized cryptographic technique into an everyday term for blockchain developers. This paper introduces the notion of ZKP for audiences without knowledge of cryptography (Sect. 2). It explores the notion of zk-SNARKs and compares them to recent alternatives, framing them in the context of blockchain (Sect. 4). In Sect. 3, I compare a few variants, with the focus on zk-SNARKs. The rest of the paper goes into more technical details, explaining how zk-SNARKs achieve their flexibility (Sect. 5) and referencing tools currently available to implement them (Sect. 6).

2 Zero-Knowledge Proofs

Zero-Knowledge Proofs were introduced in 1985 by Goldwasser, Micali and Rackoff [21]. These are a cryptographic technique in which two parties, the Prover and the Verifier, participate in a protocol. Following normal naming in the literature, I will call the Prover Peggy, and the Verifier Victor. Peggy and Victor both know a predicate

[1] See for example the recommendations in https://bitcoin.org/en/protect-your-privacy.

[2] https://z.cash.

[3] https://www.getmonero.org/.

[4] https://grin-tech.org/.

[5] https://www.beam.mw/.

[6] https://www.dash.org/.

S and a public instance x. The aim of the protocol is for Peggy to prove to Victor that $S(x)$ is a true statement without revealing anything else about why that is true. An example might be a statement about a specific graph G: "The graph G is 3-colourable". If this is true about G, Peggy can prove that is so, without revealing why it is so. That means Victor will not be able to learn any valid 3-colouring of G from Peggy's proof.

A stronger notion of Zero-Knowledge Proof is the *Zero-Knowledge Proof of Knowledge (ZK-POK)*. With such a proof, Peggy can convince Victor not only of the truth of $S(x)$ but also that Peggy knows a witness that demonstrates it. Typically, this means Peggy knows why or how the statement can be true, and this knowledge is represented by some private witness w known to Peggy but not to Victor. In this case, both Peggy and Victor know another predicate, a relation R such that $S(x)$ is true if and only if $R(x, w)$ is true as well. Continuing the example above, in a ZK-POK Peggy convinces Victor not only that $S(x)$ is true, but also that she knows w such that $R(x, w) = 1$.

2.1 *Zero-Knowledge and* **NP**

The class of predicates that can be proven in Zero-Knowledge is well defined: it coincides with the class **NP**, under the mild assumption that encryption functions exist. This class is made up of exactly those languages which can be *verified* in polynomial time, that is:

Definition 1 Language L is in **NP** if there is a relationship $R_L(x, w)$ that runs in time polynomial on the size of its input x and can verify membership in the language: if x belongs in the language, then there is a witness w related to x. If x is not in the language, then there is no witness that can be related to it. Formally,

$$\forall x \in L,\ \exists\, w\ s.t.\ R_L(x, w) = 1$$
$$\forall x \notin L,\ \forall\, w\ R_L(x, w) = 0.$$

This result was proved in [20].

An example of such a language is the non-primality of integers. Define the language *NONPRIMES* to be composed of all integers which are not prime. The relationship for this language is

$$R_{NONPRIMES}(x, s) = \ x \bmod s == 0\ \wedge$$
$$s \neq x\ \wedge s \neq 1,$$

and for each member x, we can provide a witness by showing a non-trivial factor of x.

2.2 Interactive and Non-interactive Proofs of Knowledge

The original Zero-Knowledge proofs were all examples of interactive proofs, where Peggy and Victor send messages to each other until Victor is satisfied. At each step, Victor sends a random query to Peggy that she can only answer successfully with some low probability in case the statement is false or she does not actually know a proof. If at any point Peggy is unable to successfully answer the query, then Victor knows she is cheating and rejects the proof. By repeating this questioning enough times, Victor reduces the probability that Peggy succeeds. For example, if Peggy's chance of success to any query is $\frac{1}{2}$, then the probability of succeeding ten times in a row is only $\left(\frac{1}{2}\right)^{10} = \frac{1}{2^{10}}$. Therefore, by issuing just 10 questions, Victor can have a very good assurance that Peggy is not lying.

There is a general approach to make a proof of knowledge non-interactive, called the Fiat-Shamir heuristic [16], as long as Victor's randomness is public to all parties. This is called a heuristic because it provides security in the random-oracle model only, that is, assuming that the hash function behaves as a good random function. By using this technique, Peggy can simulate the random queries Victor would pose her by replacing them with a hash of the previous message in the protocol. The first message is usually sent by Peggy, committing to a blinded version of her private input in order to guarantee that Victor does not learn anything about it (and maintain the Zero-Knowledge property), but also ensuring she cannot fool Victor by using a different value. Using this heuristic, Peggy can compute all of "Victor's messages", which she then can send in a single transcript of the whole session. Victor's work then reduces to verifying this single transcript and output whether he thinks that is a correct proof.

A different way to make a proof non-interactive is the use of a Common Reference String (CRS), proposed in [10] and expanded in [9]. This model differs from the Fiat-Shamir heuristic in that it uses true randomness, and not a simulation thereof. This is created in a setup phase and given to all participants. Some measure of trust is needed in this phase, as all participants must be assured that the string has been honestly generated and is correctly shared by all parties.

A special instance of non-interactive proofs of knowledge is known as zk-SNARKs. In fact, they're not proofs, but rather arguments of knowledge. A proof of knowledge guarantees that a malicious prover cannot prove any false statement. On the other hand, an argument of knowledge only gives such guarantees with respect to *computationally bounded provers*. The defining properties of zk-SNARKs is that they are succinct, that is, the proofs are very small (in fact, of constant size, no matter the complexity of the proof statement) and the verification is very fast. But they are constructed in the *common reference string model*, which has the drawback of requiring a trusted setup phase. zk-SNARKs were first defined in [8], initiating a very fruitful line of research. Practical zk-SNARKs based on Arithmetic Circuits and Quadratic Arithmetic Programs were introduced in [19].

Other more recent variants of Zero-knowledge non-interactive constructions are bulletproofs [12] and zk-STARKs [4]. Bulletproofs are especially suited to a specific

	TIME		SIZE		SECURITY		
	Proof	Verification	Key	Proof	Post- Quantum Resistant	Untrusted Setup	Assumptions
zk-SNARKs	n log n	O(1)	n	O(1)	No	No	Strong Number Theoretical
zk-STARKs	n polylog n	polylog n	0	log^2 n	Yes	Yes	Weak: CHRF
Bulletproofs	n log n	n log n	n	log n	No	Yes	Weak Number Theoretical

Fig. 1 Comparison between zk-SNARKs, zk-STARKs and bulletproofs

kind of proofs, demonstrating that the secret value falls within a certain range, but they can also be used for general **NP** circuits. zk-STARKs, on the other hand, are generic constructions and have a number of advantages, but the technology is still not mature enough to be usable in practice.

3 Why zk-SNARKs?

In practice, Bulletproofs, zk-SNARKs, and zk-STARKs are all interesting technologies to use. In this section, I compare them according to some relevant parameters and discuss why at the moment zk-SNARKs seem to be the most popular choice for use with blockchain technologies. These can be divided in two groups: performance and security. It will be noted that zk-STARKs are the preferred choice in terms of security, where they beat or equal the other two options. On the contrary, zk-SNARKs excel in performance when compared with the others, and zk-STARKs in particular are still eminently not practical due to their large proof sizes. As a consequence, zk-SNARKs are the preferred general-use choice at the moment, but can be overtaken by zk-STARKs if ongoing research can make them more effective. Bulletproofs hold the middle ground. They are more secure than zk-SNARKs and notably don't require a trusted setup. At the same time, they can also be performant for simple circuits, and gain from not requiring pairing technology. Still their proofs and verification size grow with the complexity of the proof statement. A summary of these aspects follows below (Fig. 1).

3.1 Performance

Proof Size The big advantage of zk-SNARKs is the proof size, that is always a constant independently of the circuit's complexity. On the other hand, the size of a bulletproof grows logarithmically with the size of the circuit. For simple statements, this is short enough to be practical, but no technology can currently beat zk-SNARKs

in the general case. zk-STARKs are particularly bad, generating proofs in the order of tens or hundreds of kB while zk-SNARKs generate proofs in the hundreds of bytes.
Verification Time Verification time is directly related to the proof size and the number of public inputs. zk-SNARKs are again very efficient in this respect, with the time growing linearly with the number of public inputs. However, the most expensive operations are a constant number of elliptic-curve pairings, that may still dominate if the public inputs are few. Expect a proof to be verified in a few milliseconds. Bulletproofs in turn can be verified in time proportional to their proof size, which is fast for simple circuits. zk-STARKs also have fast verification, but still not as fast as zk-SNARKs.

The zk-STARK whitepaper claims zk-STARKs verify in a constant time, while zk-SNARKs's time would grow linearly. However, they include the setup time in this (which does not exists for zk-STARKs). They also show a comparison when the setup is not included, which shows zk-SNARKs about 10 times faster than zk-STARKs.

I believe they make an unfair comparison. This is because the SNARK setup is performed *only once per circuit*, whereas a single proof can be verified several times. These are clearly two separate processes, and their time should not be brought together as if the setup would be needed every time we make a verification, as that is plainly not the case. Therefore, the valid comparison for me shows zk-SNARKs are faster than zk-STARKs.
Key Size zk-STARKs and Bulletproofs get the upper hand here as neither require any keys, whereas zk-SNARKs do. In particular, proving keys can be extremely large for complex circuits (in the order of megabytes of even gigabytes, depending on the circuit complexity), since they essentially encode the whole computation.

3.2 Security

Trusted Setup zk-SNARKs are constructed in the Common Reference String Model. This means they require a setup phase which creates a Common Reference String (CRS) made up of a proving key and a verification key that can be then distributed to appropriate users. This setup is highly sensitive. As part of the key generation, some randomness is created that must be destroyed at the end of the setup. Otherwise, an attacker who learns of this would be able to create false proofs. This is a difficult problem when the CRS has to be generated for use in a large network of untrusted participants.

In comparison, Bulletproofs and zk-STARKs do not require such a setup, which gives them an important advantage.
Hardness Assumptions Proofs of security usually depend on some hardness assumption: they reduce any possible attack (within a certain model) to breaking a known problem that is considered to be very hard. The weaker an assumption is, the more likely it is thought to be true and the less things it requires, the stronger the proof of security is. zk-SNARKs require very strong assumptions of number-theoretical nature, namely the relatively recent and still insufficiently understood Knowledge-

of-Exponent Assumption [3]. Bulletproofs are also based on a number-theoretical assumption, the hardness of the discrete logarithm, but this is a standard assumption and much weaker than that needed for zk-SNARKs. zk-STARKs require the weakest assumptions of all three, merely the existence of a collision-resistant hash function. *Post-Quantum Resistance* Both zk-SNARKs and Bulletproofs require number-theoretical assumptions of a nature that will be easily broken in case quantum-computers become practical. On the other hand, zk-STARKs' assumption is not number-theoretical and is currently not known to be broken by quantum-computers. Therefore, zk-STARKs are considered to be post-quantum resistant, whereas the other two types of construction are not.

3.3 Existing Snarks

Although the literature in zk-SNARKs is quite extensive, only a few of them have been implemented in practical cryptographic tools. The most popular ones are [6, 22], due to their proofs of constant size and verification time linear only in the input. Both are based on the same QAP front-end. The former is an update of the original [29], and has been available for a longer time. Although an attack was found on its definition this year [18], this has been fixed and the scheme is considered secure again. The Groth scheme is currently the most efficient one available, with shorter proofs. There is another alternative, [23], that has a proof as short as [22], but gives stronger guarantees, since it is actually a *signature of knowledge* and not just a proof of knowledge. It is, however, less efficient in both proof generation and verification.

Other earlier constructions can be found in [2, 5, 15, 19, 29].

4 Use in Blockchains

zk-SNARKs are particularly well suited to work in blockchains for two main reasons. First, they are non-interactive, which means verification can happen independently from the prover. This allows several verifiers to check the proof without collaboration and at their convenience. Secondly, the proof is concise, which means it can be conveniently given to a smart-contract without incurring a heavy gas cost and making the verification fast as well. But the main reason they are interesting is because of the functionality they bring, which is crucial for blockchain use-cases: privacy and scalability.

The case for privacy has already been argued at the beginning of this paper: I believe it is crucial for high mass-adoption in use cases where users require privacy, be it for their financial or commercial data. zk-SNARKs provide the ability to hide all of these from the public. It is possible to make the blockchain store only summaries of masked versions of a state, and enforce the consistency of updates by Zero-

Knowledge proofs. As an example, consider the case of a transaction in the UTXO model: this will include a list of all input notes, which together represent the money spent; and a list of output notes, which represent the money sent to the recipient(s). In Bitcoin, these notes are represented in the clear, but a private alternative could send just the hashes of these notes. This would totally obfuscate the balances spent and received and would even make possible the creation and destruction of value. In order to enforce consistency, we can add a Zero-Knowledge proof that the sum of inputs and outputs is the same (or differs only by the network fees) without revealing them. This is the approach taken by ZCash [31].

Privacy in this way immediately gives rise to scalability improvements. Since a single transaction can now be summarized by a short state update and a short proof, and because this proof can be automatically verified, then we can also do a single proof that validates other proofs. Instead of submitting a zero-knowledge proof for each transaction, we can effectively bundle a group of them, providing proofs for each, and then add a single proof that checks all of them. Only this proof is submitted to the chain, together with an update of state that reflects all of the verified transactions. An example of a project that is exploring this approach is CODA (see [27]), which builds on the academic work of [7].

5 QAP-Based Snarks

QAP-based zk-SNARKs have become popular for practical use-cases because of their flexibility. Once the predicate statement is codified in terms of an arithmetic circuit, a zk-SNARK can automatically be built by appropriate tools that turn the circuit into a QAP and then use that to setup the system. The implementer's job is mostly focused on specifying the R1CS that defines the problem, which can be done more or less trivially once the circuit is defined (the non-trivial aspect is that there may be thousands of gates and wires in the circuit).

Compare, for example, with $\Sigma-$protocols, which can still be used for a large variety of proofs but where the designer has to be much more careful in deciding the content of each message. Some results show how we can mix and match basic protocols to prove more complex statements (see for example [1, 13, 14, 26]), but as far as I know there is not a simple compact way to encode an arbitrary **NP** statement into a single proof.

For that reason, in this section I focus on the progression from arithmetic circuits to QAPs, explaining how these effectively encode the whole computation and so make the proof convincing.

5.1 Arithmetic Circuits

Circuits have long been used as a computation model. A circuit is composed of wires, which carry values, and gates, which perform an operation on their input values and return one or more outputs. In complexity theory, it is common to consider logical circuits, where the value of each wire is either 0 or 1, and gates perform logical operations like **AND**, **OR**, **NOT** and their variants. For zk-SNARKs, we use instead arithmetic circuits. The only difference to logical circuits is the operations computed by their gates and the values that each wire can represent.

The gates of a circuit must be organized in a directed acyclic graph, so that computation always flows in a single direction from inputs to outputs. Unlike a programming language, circuits do not have loops or functions where the computation can return multiple times to the same place: computations are flattened and laid out in a way that each gate is evaluated only once and the wires, once set, never change value. Analogously to simple electric circuits, where you can determine the voltage and current of every wire once you turn the power on, you can 'instantaneously' determine the value of all wires, including the output, of a logical circuit once you set the input values.

QAP-Based Snarks are based on pairings over elliptic curves, which are ultimately used to encode all the steps of the computation. The arithmetic circuit used for a zk-SNARK is tied to the specific finite field underlying the elliptic curve used, and so each wire can represent a single field element. The circuit is the verifier of a computation, and so returns a binary value. For a valid proof, the circuit should return 1 if the private input (the witness) provided by the prover matches the public input known to both parties.

Gates can perform modular addition and multiplication. Note that the modulus used is the order of the curve, and not that of the field. This is because of how the circuit is encoded to produce the zk-SNARK: the values of each wire are multiplied with curve points combined in several linear combinations. In other words, they are the scalars resulting from a series of multiplications in the elliptic curve's group. Therefore, they will never be larger than the curve's order.

Figure 2 demonstrates a simple arithmetic circuit, and how the computation proceeds from the input wires to produce values to the outputs. With this circuit, the prover demonstrates knowledge of two private values, a and b, that satisfy a certain relationship with a public input, n, namely $a^2 + ab - b = n$.

Each wire is thus assigned a single value, and the list of all these assignments constitutes an instance of the circuit's computation, which corresponds to a single input. Conventionally, in an assignment public wires are listed first, followed by private inputs and then the internal wires corresponding to intermediate computations. Circuit outputs are always public. The whole is preceded by a constant 1 that is added to enable constant values. The example in the figure corresponds to the following assignment, with one public input, one (public) output, two private inputs and four internal wires:

$$[1, 9, 1, 2, 5, 4, 10, 14, 14]. \tag{1}$$

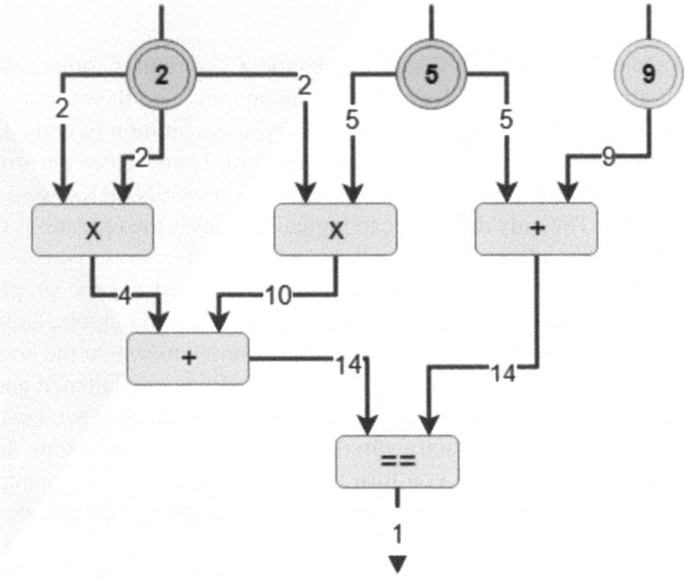

Fig. 2 A simple arithmetic circuit proving knowledge of a, b such that $a^2 + ab - b = n$

If we give a different input to the circuit, say $a = 4, b = 3, n = 10$, the result will be a failing computation, and the full assignment will be

$$[1, 10, 0, 4, 3, 16, 12, 28, 13].$$

5.2 Rank-1 Constraint Systems

An assignment of values to all the wires in the circuit describes a single computation for a given set of public and private inputs. The next step in the construction of a zk-SNARK is to produce a set of constraints that assert this computation has been correctly performed, and that the assignment is internally consistent. In other words, the constraints check that each non-input wire follows correctly from the application of a gate operation to the input's gates. Therefore, we create a rank-1 constraint for each gate, and call Rank-1 Constraint System (R1CS) to the set of all constraints. Consequently, if a cheating prover does not know the correct private witness for the public input and tries to fool the verifier by showing an output that does not follow from its inputs, then at some point a gate's output must have been miscalculated and the corresponding constraint will not be valid.

Constraints are encoded as a simple multiplicative form:

$$a \times b = c,$$

where a, b and c are numbers resulting from evaluating linear combinations of wires (A, B and C), as will be seen below. Given the circuit above, and a well-defined wire ordering, we can encode the first multiplication gate, which computes a^2, by

$$\begin{aligned}
A &= [0, 0, 0, 1, 0, 0, 0, 0, 0] \text{ representing the first private input} \\
B &= [0, 0, 0, 1, 0, 0, 0, 0, 0] \text{ representing the same wire} \\
C &= [0, 0, 0, 0, 0, 1, 0, 0, 0] \text{ representing the first internal wire.}
\end{aligned}$$

An addition gate must be represented by adding wires within the same linear combination. Consequently, the other input linear combination must simply represent 1. This is the representation of the first addition gate:

$$\begin{aligned}
A &= [0, 0, 0, 0, 0, 1, 1, 0, 0] \text{ representing the addition of two internal wires} \qquad (2) \\
B &= [1, 0, 0, 0, 0, 0, 0, 0, 0] \text{ representing the constant 1 wire} \\
C &= [0, 0, 0, 0, 0, 0, 0, 1, 0] \text{ representing the third internal wire.}
\end{aligned}$$

The constraint system may include more constraints than just those implied by the gates. As an example, equality constraints can be added. An equality constraint on two wires can be written as a multiplication: one of the inputs is set to 1, and the other two encode the wires that are being compared.

More complex assertions can be composed in this form. For example, the constraint wire 3 can carry only a binary value, which can be described by the equation $w_3 \cdot (w_3 - 1) = 0$ or equivalently $w_3^2 = w_3$. Such a constraint would not ordinarily be represented in the circuit, since it does not have an impact in the computation, but should be added to the R1CS.

All of the matrices above simply encode an abstract verification. Their concrete meaning is given by mixing them with a wire-assignment corresponding to a computation. Let such assignment be a vector $\vec{s}$. Then, a constraint (A, B, C) is satisfied for the computation represented by $\vec{s}$ if and only if

$$\langle A \cdot \vec{s} \rangle \times \langle B \cdot \vec{s} \rangle = \langle C \cdot \vec{s} \rangle.$$

For example, taking the constraint in (2) and the assignment in (1), we have

$$\begin{aligned}
a &= \langle [0, 0, 0, 0, 0, 1, 1, 0, 0] \cdot [1, 9, 1, 2, 5, 4, 10, 14, 14] \rangle = 14 \\
b &= \langle [1, 0, 0, 0, 0, 0, 0, 0, 0] \cdot [1, 9, 1, 2, 5, 4, 10, 14, 14] \rangle = 1 \\
c &= \langle [0, 0, 0, 0, 0, 0, 0, 1, 0] \cdot [1, 9, 1, 2, 5, 4, 10, 14, 14] \rangle = 14.
\end{aligned}$$

Finally, it is worth noting that constraints can be more involved than the simple examples here, and that coefficients are often larger than 1.

5.3 Polynomial-Encoding

In practical circuits, there can be thousands or millions of wires and constraints, so that verifying them all individually would be too expensive. Instead, zk-SNARKs proceed by encoding all the constraints into three polynomial vectors, which allows for the simultaneous verification of the whole constraint set.

Let n represent the number of wires in an assignment, and k the number of constraints. Denote these by

$$\mathbb{C}_1 = (\mathbf{A}_1, \mathbf{B}_1, \mathbf{C}_1)$$
$$\ldots\ldots$$
$$\mathbb{C}_k = (\mathbf{A}_k, \mathbf{B}_k, \mathbf{C}_n)$$

Let

$$[\mathbf{A}] = \begin{bmatrix} \mathbf{A}_1[1] \; \mathbf{A}_1[2] \cdots \mathbf{A}_1[n] \\ \cdots \\ \mathbf{A}_k[1] \; \mathbf{A}_k[2] \cdots \mathbf{A}_k[n] \end{bmatrix}$$

be the matrix of all A linear combinations.

We can devise a polynomial vector $\mathcal{A}$ with n elements that describes $[\mathbf{A}]$. Each member $\mathcal{A}_i$ of $\vec{\mathcal{A}}$ is a polynomial that encodes the ith column of $[\mathbf{A}]$. We assign to each constraint $\mathbb{C}_j$ a fixed scalar value, σ_j. The pairs $(\sigma_1, \mathbf{A}_1[i]), \ldots, (\sigma_k, \mathbf{A}_k[i])$ represent the ith coordinate of all the constraints as points on a plane. Now define $\mathcal{A}_i$ as a polynomial that passes through these k points, for example as the result of the Lagrange interpolation. Define analogously $\vec{\mathcal{B}}$ and $\vec{\mathcal{C}}$ for the other linear combinations. Next, we gather the terms of all polynomials evaluated at the coordinate for constraint j under the following notation:

$$\vec{\mathcal{A}}\,(\sigma_j) = [\mathcal{A}_1(\sigma_j), \mathcal{A}_2(\sigma_j), \ldots, \mathcal{A}_n(\sigma_j)] \tag{3}$$

and observe that by construction this is

$$\vec{\mathcal{A}}\,(\sigma_j) = [\mathbf{A}][j] = \mathbf{A}_j.$$

It can now be easily checked that, for an assignment $\vec{s}$ as above and an arbitrary constraint $\mathbb{C}_j$, with $j \in \{1, \ldots, k\}$, and for all $i \in \{1, \ldots, n\}$:

$$\langle \vec{\mathcal{A}}(\sigma_j) \cdot \vec{s} \rangle \cdot \langle \vec{\mathcal{B}}(\sigma_j) \cdot \vec{s} \rangle = \langle \vec{\mathcal{C}}(\sigma_j) \cdot \vec{s} \rangle \Leftrightarrow$$
$$\langle \vec{\mathcal{A}}(\sigma_j) \cdot \vec{s} \rangle \cdot \langle \vec{\mathcal{B}}(\sigma_j) \cdot \vec{s} \rangle - \langle \vec{\mathcal{C}}(\sigma_j) \cdot \vec{s} \rangle = 0 \Leftrightarrow$$
$$\langle \mathbf{A}_j \cdot \vec{s} \rangle \cdot \langle \mathbf{B}_j \cdot \vec{s} \rangle - \langle \mathbf{C}_j \cdot \vec{s} \rangle = 0.$$

is satisfied if and only if the computation satisfies the jth constraint.

The above expression can be further developed, to show that it represents a simple polynomial expression of the kind $\mathcal{P}(\sigma) = 0$:

$$\mathcal{P}(\sigma) = \left(\sum_{i=1}^{n} \mathcal{A}_i(\sigma) \cdot s_i\right) \cdot \left(\sum_{i=1}^{n} \mathcal{B}_i(\sigma) \cdot s_i\right) - \left(\sum_{i=1}^{n} \mathcal{C}_i(\sigma) \cdot s_i\right) = 0 \quad (4)$$

5.4 *Proof Construction*

The computation verified by a zk-SNARK should be known by both the Prover and the Verifier, and therefore the set of constraints and the corresponding polynomial vector will also be known by both. These set the rules of the computation. Recall that the proof asserts knowledge of a witness $\vec{s}$ that passes those constraints. This section details how the Prover can convince the Verifier of that.

I focus on the polynomial (4) developed in the last section. If the computation satisfies all constraints, then by construction $\mathcal{P}(\sigma) = 0$ at least when $\sigma \in S = \{\sigma_1, \ldots, \sigma_k\}$ and possibly at other points. It is important to notice here that $\mathcal{P}(\sigma)$ is defined for a specific witness $\vec{s}$ and so encodes a specific computation.

By a consequence of the fundamental theorem of algebra, $\mathcal{P}(\sigma)$ must be a multiple of a polynomial $\mathcal{Z}(\sigma)$ that vanishes exactly in set S, that is, $\mathcal{Z}(\sigma) = 0 \Leftrightarrow \sigma \in S$. This polynomial is defined as:

$$\mathcal{Z}(\sigma) = (\sigma - \sigma_1) \cdot (\sigma - \sigma_2) \cdot \ldots \cdot (\sigma - \sigma_k).$$

The reverse is also true, that is, if $\mathcal{Z}(\sigma)$ evenly divides $\mathcal{P}(\sigma)$, then $\mathcal{P}(\sigma)$ must vanish on set S and so all constraints are satisfied by the witness $\vec{s}$. Therefore, to prove the correctness of a computation, it is enough to demonstrate that $\mathcal{P}(\sigma)$ is a multiple of $\mathcal{Z}(\sigma)$ by showing $\mathcal{H}(\sigma)$ such that $\mathcal{P}(\sigma) = \mathcal{H}(\sigma) \cdot \mathcal{Z}(\sigma)$.

Notice that the Verifier can compute $\mathcal{Z}(\sigma)$, and that this determines what a Prover needs to compute in order to produce a convincing proof. Thus, the whole computation can be specified in a Quadratic Arithmetic Program composed of

$$\text{QAP} = (\vec{\mathcal{A}}, \vec{\mathcal{B}}, \vec{\mathcal{C}}, \mathcal{Z}). \quad (5)$$

The polynomial $\mathcal{H}(\sigma)$ and the witness $\overrightarrow{s}$ are the heart of the proof. Although the verifier can not compute $\mathcal{P}(\sigma)$, because it does not know $\overrightarrow{s}$, it knows it must follow from QAP and that this is embedded in the proving and verification keys in a way that only a valid $\mathcal{H}$ will make the proof valid.

This encoding is the basis for QAP-based zk-SNARKs. They share the same 'front-end', by encoding the computation into a QAP in the same way. It is in the subsequent constructions, how that computation is encoded and how the proof is created, that they differ. That is not in the scope for this paper, and I encourage the reader to consult the references in Sect. 3.3.

6 Tools

Currently, there are a limited number of tools supporting the development of zk-SNARK applications. All of those I know focus mainly on QAP-based zk-SNARKs. These tools can be divided in 2 layers: Domain Specific Languages (DSL) that allow describing the proof predicate in a high-level language that is easy to learn and use; and support for the construction of the zk-SNARK algorithms, including (i) the representation of the statement in some technical intermediate language; (ii) support for the zk-SNARK specific algorithms (CRS generation, proof creation and verification); (iii) all the necessary mathematical support, for fast calculation in very large fields, elliptic curves and bilinear pairings. Most of these use R1CS as the intermediate language, which is then compiled into a QAP. R1CS is close to the language of arithmetic circuit satisfiability and therefore is **NP**-complete. For this reason, it has become very popular and is used by all the libraries reviewed here.

6.1 Zk-SNARK Support Libraries

The most complete library to the date is also one of the first: libsnark.[7] It is written in C++ and offers wide flexibility. It mainly uses R1CS as the language for representing the proof predicate, but can support other more efficient (but possibly less flexible) languages, such as BACS, USCS, TBCS. libsnark offers also different kinds of elliptic curves, via its dependency libff,[8] supporting BN, Edwards and MNT curves. It also supports different kinds of Snarks, including BCTV14, Groth16 and GM17.

DIZK[9] is another library published by SCIPR Labs, and is in some aspects like a reduced port of libsnark to Java. Instead of relying on external libraries for calculation of FFT and fast arithmetic, it integrates its own implementations for these tasks, but in some cases (eg FFT) with a reduced algorithm. DIZK's main selling point is the

[7] https://github.com/scipr-lab/libsnark.

[8] https://github.com/scipr-lab/libff.

[9] https://github.com/scipr-lab/dizk.

support for parallelization to speed up the setup and proof generation. It supports only 2 BN curves, and offers only one kind of zk-SNARK, Groth16, which is QAP-based and described in the R1CS language.

Snarkjs[10] is a library for implementing zk-SNARKs in Javascript. Again it is limited to only one specific type of curve (BN128), and implements two Snarks, the original BCTV14 and Groth16. Due to the narrow elliptic curve focus, the library is smaller and probably easier to understand than either DIZK or libsnark.

Bellman[11] is a compact library, being developed for ZCash, that supports the creation of Groth16 Snarks. It is developed in Rust, and supports only Groth16, again based on an R1CS representation. Unlike the previous systems, it seems to use only the BLS12-381 curve, as promoted by ZCash since the Sapling version.

6.2 DSL Tools

Typically, zk-SNARK libraries are difficult to use without a way to encode an arbitrary proof predicate. This niche is covered by some dedicated libraries.

ZoKrates[12] is a tool written in Rust and C++ that offers a very simple language to encode arithmetic circuits and R1CS. It interfaces with libsnark and Bellman, and allows the creation of three types of zk-SNARK: BCTV14, GM17 and since recently Groth16. It is very actively developed, and a good choice for beginners.

jSnark[13]/xjSnark[14] are a pair of libraries in Java that simplify the specification of zk-SNARKs at a high-level. Its approach is quite different of ZoKrates, in that the language is more low-level, centered around the definition of gadgets. But this makes it more flexible than ZoKrates, since the programmer can fine tune the construction of the circuit. Despite being written primarily in Java, its default backend is libsnark.

Circom[15] is the complement of Snarkjs. Also written in JavaScript, it provides a language that has similarities to ZoKrates, but in a more C-like way. Its intended backend is Snarkjs.

Finally, Snarky[16] is an OCaml front-end for creating R1CS-based Snarks. It uses the libsnark backend by default, and differs from the other DSL in that its language has a functional approach.

---

[10]https://github.com/iden3/snarkjs.

[11]https://github.com/zcash/librustzcash/tree/master/bellman.

[12]https://github.com/Zokrates/ZoKrates.

[13]https://github.com/akosba/jsnark.

[14]https://github.com/akosba/xjsnark.

[15]https://github.com/iden3/circom.

[16]https://github.com/o1-labs/snarky.

References

1. Agrawal, S., Ganesh, C., Mohassel, P.: Non-interactive zero-knowledge proofs for composite statements. In: Shacham, H., Boldyreva, A. (eds.) Advances in Cryptology - CRYPTO 2018, pp. 643–673. Springer International Publishing, Cham (2018)
2. Backes, M., Barbosa, M., Fiore, D., et al.: Adsnark: nearly-practical privacy-preserving proofs on authenticated data. In: Proceedings of the 36th IEEE Symposium on Security and Privacy (S&P), May 2015
3. Bellare, M., Palacio, A.: The knowledge-of-exponent assumptions and 3-round zero-knowledge protocols. In: Proceedings of Advances in Cryptology - CRYPTO 2004, 24th Annual International Cryptology Conference, Santa Barbara, California, USA, 15–19 August 2004. Lecture Notes in Computer Science, vol. 3152, pp. 273–289. Springer (2004)
4. Ben-Sasson, E., Bentov, I., Horesh, Y., et al.: Scalable, transparent, and post-quantum secure computational integrity. IACR Cryptol. ePrint Archi. **2018**, 46 (2018)
5. Ben-Sasson, E., Chiesa, A., Genkin, D., et al.: Snarks for c: verifying program executions succinctly and in zero knowledge. In: Canetti, R., Garay, J.A. (eds.) CRYPTO (2). Lecture Notes in Computer Science, vol. 8043, pp. 90–108. Springer (2013)
6. Ben-Sasson, E., Chiesa, A., Tromer, E., et al.: Succinct non-interactive zero knowledge for a von neumann architecture. In: Proceedings of the 23rd USENIX Conference on Security Symposium. SEC'14, pp. 781–796 (2014)
7. Ben-Sasson, E., Chiesa, A., Tromer, E., et al.: Scalable zero knowledge via cycles of elliptic curves. Algorithmica **79**(4), 1102–1160 (2017). https://doi.org/10.1007/s00453-016-0221-0
8. Bitansky, N., Canetti, R., Chiesa, A., et al.: From extractable collision resistance to succinct non-interactive arguments of knowledge, and back again. In: Proceedings of the 3rd Innovations in Theoretical Computer Science Conference. ITCS '12, pp. 326–349. ACM, New York (2012)
9. Blum, M., De Santis, A., Micali, S., et al.: Noninteractive zero-knowledge. SIAM J. Comput. **20**(6), 1084–1118 (1991)
10. Blum, M., Feldman, P., Micali, S.: Non-interactive zero-knowledge and its applications. In: Proceedings of the Twentieth Annual ACM Symposium on Theory of Computing. STOC '88, pp. 103–112. ACM, New York (1988)
11. Bünz, B., Agrawal, S., Zamani, M., et al.: Zether: towards privacy in a smart contract world. IACR Cryptol. ePrint Arch. **2019**, 191 (2019). https://eprint.iacr.org/2019/191
12. Bünz, B., Bootle, J., Boneh, D., et al.: Bulletproofs: short proofs for confidential transactions and more. In: Proceedings of 2018 IEEE Symposium on Security and Privacy, SP 2018, San Francisco, California, USA, 21–23 May 2018, pp. 315–334 (2018)
13. Ciampi, M., Persiano, G., Scafuro, A., et al.: Improved or-composition of sigma-protocols. In: Proceedings of Theory of Cryptography - 13th International Conference, TCC 2016-A, Tel Aviv, Israel, Part II, 10–13 January 2016, pp. 112–141 (2016). https://doi.org/10.1007/978-3-662-49099-0_5,
14. Cramer, R., Damgård, I., Schoenmakers, B.: Proofs of partial knowledge and simplified design of witness hiding protocols. In: Proceedings of the 14th Annual International Cryptology Conference on Advances in Cryptology. CRYPTO '94, pp. 174–187. Springer, London (1994). http://dl.acm.org/citation.cfm?id=646759.705842
15. Danezis, G., Fournet, C., Groth, J., et al.: Square span programs with applications to succinct NIZK arguments. In: ASIACRYPT (1). Lecture Notes in Computer Science, vol. 8873, pp. 532–550. Springer (2014)
16. Fiat, A., Shamir, A.: How to prove yourself: practical solutions to identification and signature problems. In: Proceedings on Advances in Cryptology—CRYPTO '86, pp. 186–194. Springer, London (1987)
17. Frankenfield, J.: Coinjoin, July 2018. https://www.investopedia.com/terms/c/coinjoin.asp. Accessed 27 May 2019
18. Gabizon, A.: On the security of the BCTV pinocchio zk-snark variant. IACR Cryptol. ePrint Arch. **2019**, 119 (2019)

19. Gennaro, R., Gentry, C., Parno, B., et al.: Quadratic span programs and succinct nizks without pcps. In: Proceedings of Advances in Cryptology - EUROCRYPT 2013, 32nd Annual International Conference on the Theory and Applications of Cryptographic Techniques, Athens, Greece, 26–30 May 2013, pp. 626–645 (2013)
20. Goldreich, O., Micali, S., Wigderson, A.: Proofs that yield nothing but their validity or all languages in np have zero-knowledge proof systems. J. ACM **38**(3), 690–728 (1991)
21. Goldwasser, S., Micali, S., Rackoff, C.: The knowledge complexity of interactive proof-systems. In: Proceedings of the Seventeenth Annual ACM Symposium on Theory of Computing. STOC '85, pp. 291–304. ACM, New York (1985)
22. Groth, J.: On the size of pairing-based non-interactive arguments. In: EUROCRYPT (2). Lecture Notes in Computer Science, vol. 9666, pp. 305–326. Springer (2016)
23. Groth, J., Maller, M.: Snarky signatures: minimal signatures of knowledge from simulation-extractable snarks. In: CRYPTO (2). Lecture Notes in Computer Science, vol. 10402, pp. 581–612. Springer (2017)
24. Jedusor, T.E.: Mimblewimble (2016). https://download.wpsoftware.net/bitcoin/wizardry/mimblewimble.txt. Accessed 27 May 2019
25. Lamport, L., Shostak, R., Pease, M.: The byzantine generals problem. ACM Trans. Program. Lang. Syst. **4**(3), 382–401 (1982)
26. Maurer, U.: Zero-knowledge proofs of knowledge for group homomorphisms. Des. Codes Cryptogr. **77**(2–3), 663–676 (2015). https://doi.org/10.1007/s10623-015-0103-5
27. Meckler, I., Shapiro, E.: Coda: decentralized cryptocurrency at scale (2018). https://cdn.codaprotocol.com/v2/static/coda-whitepaper-05-10-2018-0.pdf. Accessed 30 May 2019
28. Nakamoto, S.: Bitcoin: a peer-to-peer electronic cash system (2008). https://bitcoin.org/bitcoin.pdf. Accessed 27 May 2019
29. Parno, B., Howell, J., Gentry, C., et al.: Pinocchio: nearly practical verifiable computation. In: IEEE Symposium on Security and Privacy, pp. 238–252. IEEE Computer Society (2013)
30. Poelstra, A.: Mimblewimble (2016). https://download.wpsoftware.net/bitcoin/wizardry/mimblewimble.pdf. Accessed 27 May 2019
31. Sasson, E.B., Chiesa, A., Garman, C., et al.: Zerocash: decentralized anonymous payments from bitcoin. In: 2014 IEEE Symposium on Security and Privacy, May 2014, pp. 459–474

The manufacturer's authorised representative in the EU is Springer Nature Customer Service Centre GmbH, Europaplatz 3, 69115 Heidelberg, Germany. If you have any concerns regarding our products, please contact ProductSafety@springernature.com

Printed and bound by CPI Group (UK) Ltd, Croydon, CR0 4YY
30/09/2026
02225458-0011